Seventh Edition
Laboratory Anatomy of the
Cat

Robert B. Chiasson
University of Arizona
Illustrated by the author
Ernest S. Booth

wcb

Wm. C. Brown Company Publishers

Dubuque, Iowa

Booth Laboratory Anatomy Series

Consulting Editors
Ernest S. Booth
Robert B. Chiasson

Laboratory Anatomy of the Cat
Robert B. Chiasson
Ernest S. Booth

Laboratory Anatomy of the Frog
Raymond A. Underhill

Laboratory Anatomy of the Human Body
Bernard B. Butterworth

Laboratory Anatomy of the Mink
David Klingener

Laboratory Anatomy of the Perch
Robert B. Chiasson

Laboratory Anatomy of the Fetal Pig
Theron O. Odlaug

Laboratory Anatomy of the Pigeon
Robert B. Chiasson

Laboratory Anatomy of the Rabbit
Charles A. McLaughlin
Robert B. Chiasson

Laboratory Anatomy of the Rat
Robert B. Chiasson

Laboratory Anatomy of the Shark
Laurence M. Ashley

Laboratory Anatomy of the Turtle
Laurence M. Ashley

Copyright © 1944, 1946, 1948 by Ernest S. Booth

Copyright © 1967, 1972, 1977, 1982 by Wm. C. Brown Company Publishers

ISBN 0–697–04722–9

All rights reserved. No part of this publication may be reproduced, stored in a retrieval system, or transmitted, in any form or by any means, electronic, mechanical, photocopying, recording, or otherwise, without the prior written permission of the publisher.

Printed in the United States of America

Contents

Preface, v

Introduction, 1

1 External Anatomy and Skin, 3

2 Skeletal System, 6

3 Muscular System, 21

4 Coelomic Membranes and Viscera, 60

5 Digestive System, 63

6 Respiratory System, 70

7 Circulatory System, 72

8 Urinary and Reproductive Organs, 86

9 Nervous System, 90

Appendix A. Size and Age Relationships of the Cat, 104

   I. Age Determination by Skeletal and Dental Characters, 104

   II. Size Relationships, 104

Appendix B. References in Related Topics, 106

Preface

This laboratory manual was first published privately by Dr. Ernest S. Booth in 1944. At that time Dr. Booth was Professor of Biology at Walla Walla College, College Place, Washington. Second (1946) and third (1948) editions were published by the Wm. C. Brown Company Publishers. The fourth, fifth and sixth editions were revised by Dr. Robert B. Chiasson.

The past six editions of this manual attest to the fact that no excuse is necessary for the publication of a laboratory guide to the anatomy of the cat. The cat is easily the most popular specimen for mammalian dissection in the United States. It is used not only as an anatomical representative of the class Mammalia but also as a substitute for human dissection when human cadavers are unavailable or psychologically distasteful. The cat's size makes it a convenient research animal and the use of this animal in physiology laboratories has increased considerably in recent years. In particular, the domestic cat has served as an experimental animal in physiological studies of muscle contraction, respiration, endocrine glands, and of the central nervous system.

Robert B. Chiasson

Introduction

The cat is a mammal, a member of the phylum Chordata, subphylum Vertebrata, class Mammalia, order Carnivora, family Felidae, genus *Felis*, and species *domestica*. Its scientific name is *Felis domestica*.

Before beginning the dissection of the cat it would be well to become familiar with a few of the general principles of dissection. In the dissection of any animal, care should be taken so that delicate structures will not be mutilated during the first stages of the work. Proper care should be used in each step of the process.

1. Be sure you know what you are cutting before you cut. Read directions carefully before dissecting any part of the animal. The dissecting probe is the most important tool you have; use it freely and the scalpel sparingly.
2. Do not ask questions that can be answered by reading directions. If you have not been able to find the structure after diligent effort, then call on the laboratory instructor for help.
3. There is some variation in the arrangement of the organs of any animal, especially of the veins. The drawings represent the usual conditions to be found in the cat, but you are sure to find exceptions. If you do find errors in this manual, we would appreciate being apprised of them.
4. Each student should furnish the following dissecting instruments:

   Scalpel, good quality (preferably with removable blade).
   Hemostat, inexpensive as possible.
   Forceps.
   Scissors, good quality, with one blunt tip.
   Dissecting needles, six or more.
   Probe, with blunt tip.
   Disposable latex surgical gloves, box of 100.

PLANES AND DIRECTIONS

The most common directional terms used in this atlas are: *cranial, caudal, dorsal, ventral, distal,* and *medial*. These terms not only serve to orient the dissection but they are also used as parts of the names of structures.

Figure 1 will help to explain these and other directional terms and planes of the cat.

Anterior, posterior, superior and *inferior* are terms reserved for human anatomy and for certain structures in the head such as the eye and ear.

Vertebrate anatomists have previously used anterior and posterior as synonyms for cranial and caudal but veterinary anatomists (Nomina Anatomica Veterinaria) have rejected the usage for mammals and avian anatomists (Nomina Anatomica Avium) have rejected the terms for birds. Some textbooks continue to use anterior and posterior to correspond to cranial and caudal but this leads to confusion with human anatomy. Human anatomists use anterior as vertebrate anatomists use ventral and posterior as vertebrate anatomists use dorsal. If comparisons are to be made between the cat and human, a good medical dictionary should be available in the laboratory. Any one of the following is recommended:

Dorland's illustrated medical dictionary, 23rd. ed. (1957) or 24th ed. (1965). Philadelphia and London: W. B. Saunders Company.
Stedman's medical dictionary, 21st. ed. (1966). Baltimore: Williams & Wilkins Co.
Blakiston's new Gould medical dictionary, (any recent edition). New York: McGraw-Hill Book Co.

The following references deal with the gross anatomy of the cat:

Crouch, J. E. 1969. *Atlas of cat anatomy*. Philadelphia: Lea & Febiger.
Davison, A., and Stromsten, F. A. 1947. *Mammalian anatomy* (with special reference to the cat), 7th ed. Philadelphia: Blakiston Co.

Field, H. E., and Taylor, M. E. 1954. *An atlas of cat anatomy*. Chicago: University of Chicago Press.

Gilbert, S. G. 1968. *Pictorial anatomy of the cat*. Seattle: University of Washington Press.

McClure, R. C., Dallman, M. J., and Garrett, P. G. 1973. *Cat anatomy: an atlas, text, and dissection guide*. Philadelphia: Lea & Febiger.

Reighard, Jacob, and Jennings, H. S. 1935. *Anatomy of the cat*, 3rd. ed. New York: Henry Holt and Co.

The following are related but not directly applicable:

Ormrod, Arthur N. 1966. *Surgery of the dog and cat: A practical guide*. Baltimore: Williams & Wilkins Co. London: Brailliere, Tindall, and Cassell.

Wilkinson, G. T. 1966. *Diseases of the cat*. New York: Pergamon Press.

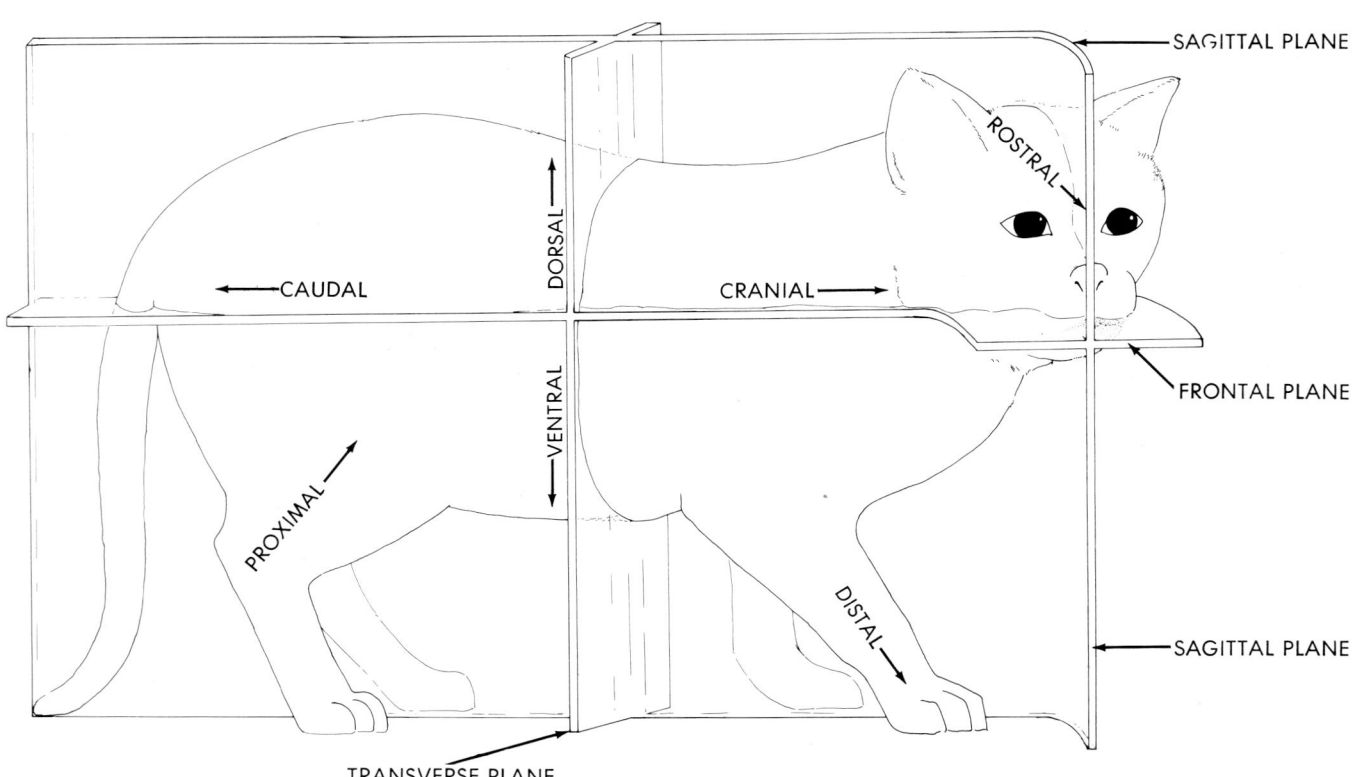

FIGURE 1. Diagram of anatomical planes and directions for the cat.

Chapter 1
External Anatomy and Skin

EXTERNAL ANATOMY

Generally speaking, orders of mammals are recognized rather easily by their external appearance; therefore, the external anatomy of a cat will be quite different from that of a rat and vice versa. The external features that separate mammals into orders are such characteristics as the number of digits on the feet, method of walking or other locomotion, characteristics of the teeth, etc.

Mammals have two unique external characteristics that distinguish them from all other vertebrates: (1) All mammals have *hair* at some time during their development, and (2) all female mammals possess *mammary glands* with external openings for nourishing the young.

The cat's head consists of a rostro-ventral facial region and a caudo-dorsal cranial portion. The lips around the mouth are well developed, and the upper lip is cleft in the center by a groove called the *philtrum*.

The external *nares* are situated on a naked nose. The eyes have upper and lower lids and a greatly reduced nictitating membrane called the *plica semilunaris*. The ears possess a long, flexible external fold called the *pinna*, which directs sound waves into the external auditory meatus.

The cat and many other mammals have special, long sensory hairs on the face called *vibrissae* or *pili tactiles*. The vibrissae are named according to their location; *supraorbitales* are above the eye, *infraorbitales* are below the eye, *zygomatici buccales* on the cheek, *buccales labiales-maxillaris* on the upper lip, and *buccales labiales-mandibularis* on the lower lip. The cat also has long hairs on the forefeet *(pili carpales)*, but they probably are not sensory hairs.

The trunk is divided into a chest, or *thorax*, and a belly, or *abdomen*. Teats or *nipples* (the external openings of the mammary glands) are located on the ventral surface of the trunk. The number and location of the nipples vary in different mammals. There are usually eight, four on each side, in the cat. The mammary glands are modified skin glands that produce milk.

Most mammals have separate urogenital and anal openings. The *anal opening* is located at the base of the tail dorsal to the urogenital orifice. In males, the urogenital structures consist of the penis and a double pouch, the *scrotum*, containing the testes.

The tail of mammals, including the cat, is usually well developed. However, the tail is reduced in rabbits and absent in humans. The horny claws on the digits are an additional epidermal derivative. Other mammals have nails, hoofs, or horns derived also from the epidermis. In the cat the claws are retractable.

The epidermis on the bottom of the cat's foot is thickened into pads called *tori* (see fig. 2). Five of the digits of the forefoot *(manus)* have pads, the *torus digitalis*, beneath each joint and between the distal and middle phalanx (see fig. 2). A larger pad, the *torus metacarpalis*, serves as a cushion for the joint between the metacarpals and the proximal phalanges (see fig. 2). An additional pad, the *torus carpalis* covers the accessory carpal bone.

Similarly the hind foot *(pes)* has a *torus metatarsalis* and four *tori tarsali*.

The position of the hand and foot in locomotion varies considerably, but it is usually constant within a single order of mammals. In the cat the gait is called *digitigrade*, meaning that the animals walk on the digits with the remainder of the hand and foot elevated. Humans are *plantigrade*, i.e., they walk on the entire sole of the foot. Horses and cattle walk on hoofs (modified claws) and are *unguligrade*.

THE SKIN

The principal covering of the mammalian body is hair. Hair is exclusively a mammalian structure and is present in all mammals at some time during their life.

Each hair consists of a central portion, the *medulla*, surrounded by a layer of elongated cells, the *cortex*, which in turn is covered by a surface layer of scale-like cells called the *cuticle*. These hairs are located in pits in the skin called *follicles*. The portion of the hair in the follicle is called the *root* and the exposed portion is termed the *shaft*. Other organs of the skin include the skin glands (*sebaceous* and *sudoriferous*), and *arrector pili* muscles.

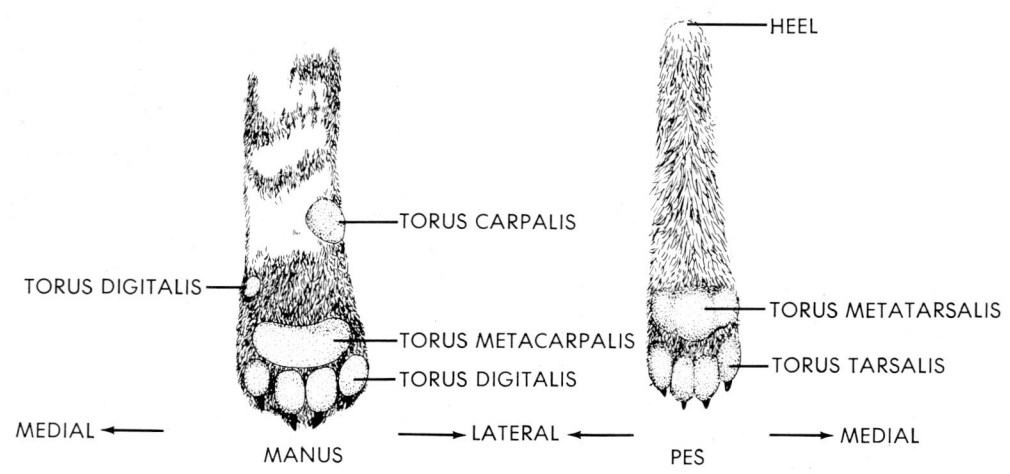

FIGURE 2. Ventral (palmar and plantar) surfaces of the right front (manus) and hind (pes) feet of the cat showing the position of *tori*.

SUGGESTED READING

Creed, R.F.S. 1958. The histology of mammalian skin, with special reference to the dog and cat. *Veterinary Record* 70:171-175.

Kuhn, R. A. 1953. Organization of tactile dermatomes in cat and monkey. *Jour. Neurophysiology* 16:169-182.

Kenshalo, D. R., Duncan, D. G., and Weymark, C. 1967. Thresholds for thermal stimulation of the inner thigh, footpad, and face of cats. *Jour. Comp. Physiol. and Psychology* 63:133-138.

Silber, I. A. 1966. The anatomy of the mammary gland of the dog and cat. *Jour. Small Animal Pract.* 7:689-696.

Strickland, J. H. 1963. The integumentary system of the cat. *Amer. Jour. Vet. Res.* 24(102):1018-1029.

Winkelmann, R. K. 1958. The sensory endings in the skin of the cat. *Jour. Comp. Neurology* 109(2):221-232.

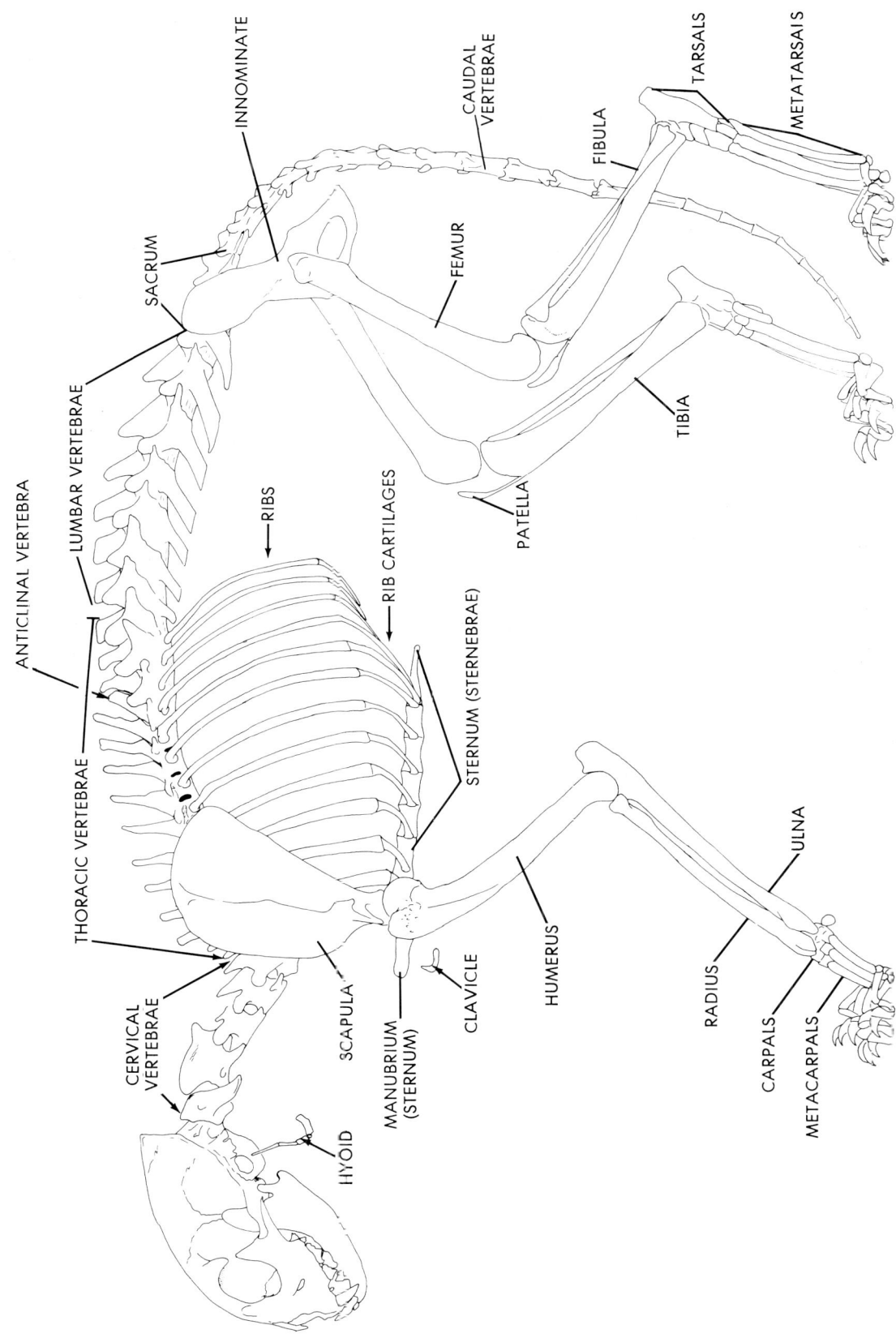

FIGURE 3. Lateral view of the cat skeleton.

Chapter 2
Skeletal System

Study both mounted and disarticulated skeletons. The mounted skeleton will show the relation of each bone to the other bones. As most of the time will be spent with the disarticulated skeleton, however, it will be necessary to learn all the bones of the body.

The skeleton can be divided into two parts: axial skeleton (bones of the skull, spinal column, and thorax), and appendicular skeleton (bones of the thoracic and pelvic girdles and appendages). Spend some time studying the entire mounted skeleton, becoming familiar with the larger bones and their relationship to each other. Spend the rest of your time studying the disarticulated skeleton.

AXIAL SKELETON

I. Skull (figs. 3-8).
   A. Cranium, caudal skull bones enclosing the brain.
      1. Frontal, paired bones with processes at the caudal border of the orbit directed toward the zygomatic arch.
      2. Parietal, paired bones. The intersection between the frontal and parietal bones is called the *bregma*.
      3. Interparietal, a small triangular bone between the two parietals and the supraoccipital.
      4. Temporals are paired bones, and each is divisible into three parts.
         a. Squamosal.
         b. Periotic *or* mastoid *or* petromastoid.
         c. Tympanic bulla.
      5. Occipital is the most caudal skull bone and is fused from four parts.
         a. Supraoccipital, the caudal wall of the cranium.
         b. Basioccipital, the caudal floor of the cranium.

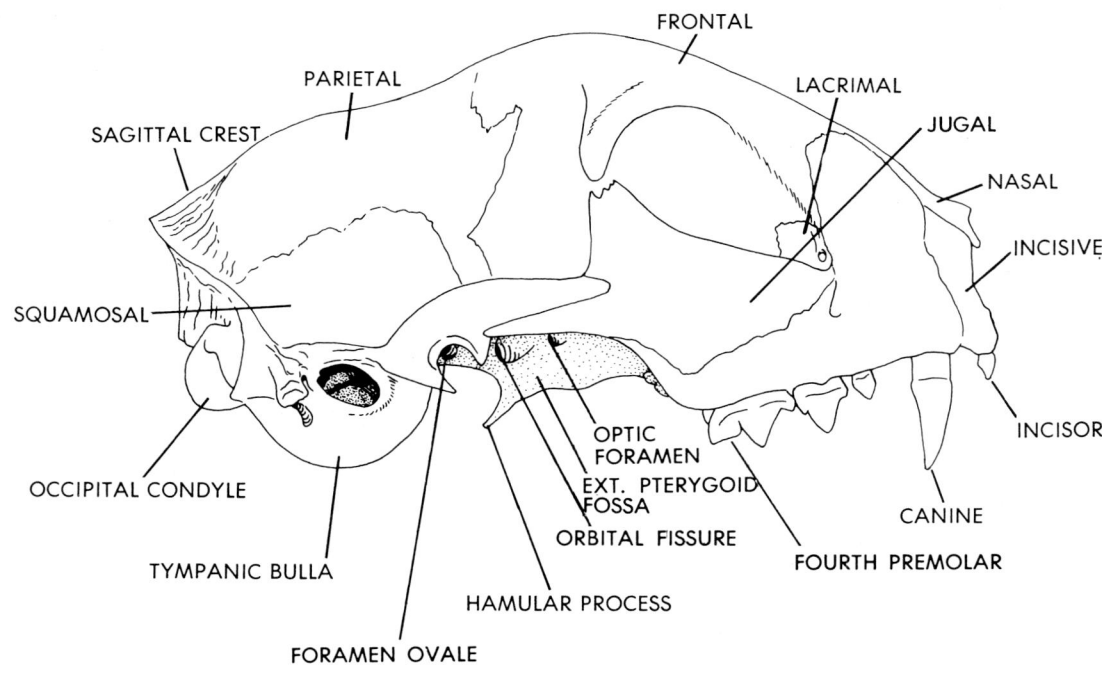

FIGURE 4. Skull, lateral view.

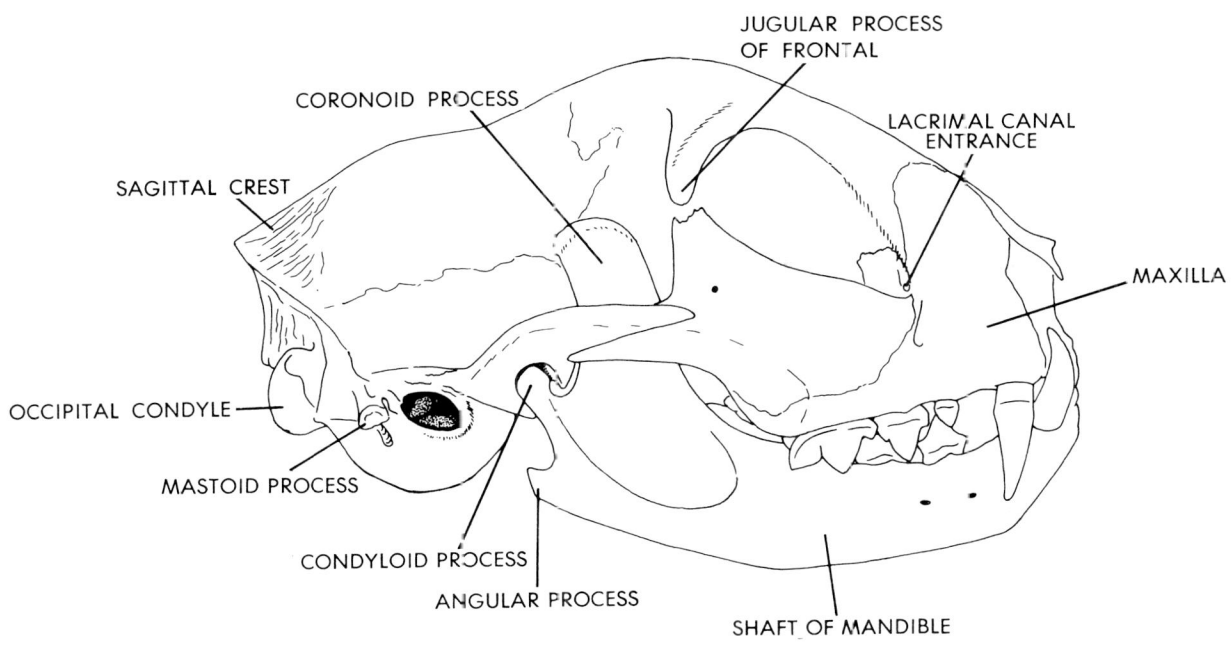

FIGURE 5. Skull and mandible, lateral view.

c. Exoccipitals, paired bones with *condyles* for articulation with the atlas.
6. Basisphenoid has two wings extending into the orbit and forms the caudal pterygoid and hanular processes. The ventrolateral (orbital) wall of the basisphenoid is the internal pterygoid fossa which serves as the origin of the internal pterygoid muscle.
7. Presphenoid, lateral portions of this bone are called *orbitosphenoids*. The external pterygoid fossa is formed by the orbital portion of this bone and the dorsolateral surface of the basisphenoid.
8. Ethmoid has a caudal perforated *cribiform plate* at the most rostral end of the cranial cavity. The remainder of this bone is a part of the splanchnocranium.

B. Facial Bones. These are bones of the splanchnocranium surrounding the digestive and respiratory systems and the special sense organs.
1. Nasals, paired bones roofing the ethmoid.
2. Incisive bones, paired bones containing the incisor teeth.
3. Maxilla, all of the upper teeth except the incisors are socketed in these paired bones.
4. Lacrimals, small paired bones at the rostral border of the orbit. The lacrimal canal passes through the dorsal rostral corner of of this bone.
5. Jugal is the central bone of the zygomatic arch. Connects the maxilla and squamosal and has a frontal process.
6. Vomer is unpaired, dorsal to the palatine bones and supports the ethmoid.
7. Ethmoid (also see no. 8 under cranium), consists of a central perpendicular plate set in the vomer (see no. 6) and thin, turbinate bones on the sides of the perpendicular plate. The caudal, perforated surface is called the cribiform plate.
8. Palatines, paired bones forming the rostral roof of the mouth (the hard palate).
9. Mandible, a paired bone that articulates with its fellow in the rostral midline by the *mandibular symphysis*. They are often fused at this point in old specimens. All of the lower teeth are socketed in the mandible. Caudally, there is a dorsal *coronoid* process for attachment of the temporalis muscle, a ventral *angular* process for attachment (medially) of the pterygoid muscles, and between these processes, a *condyloid* process for articulation with the squamosal. A single, lower jawbone (on each side) is a primary mammalian character.
10. Hyoid, consists of eleven bones, ten paired units and an unpaired *basihyal*. The cranial horn has four pairs of bones: a *ceratohyal*

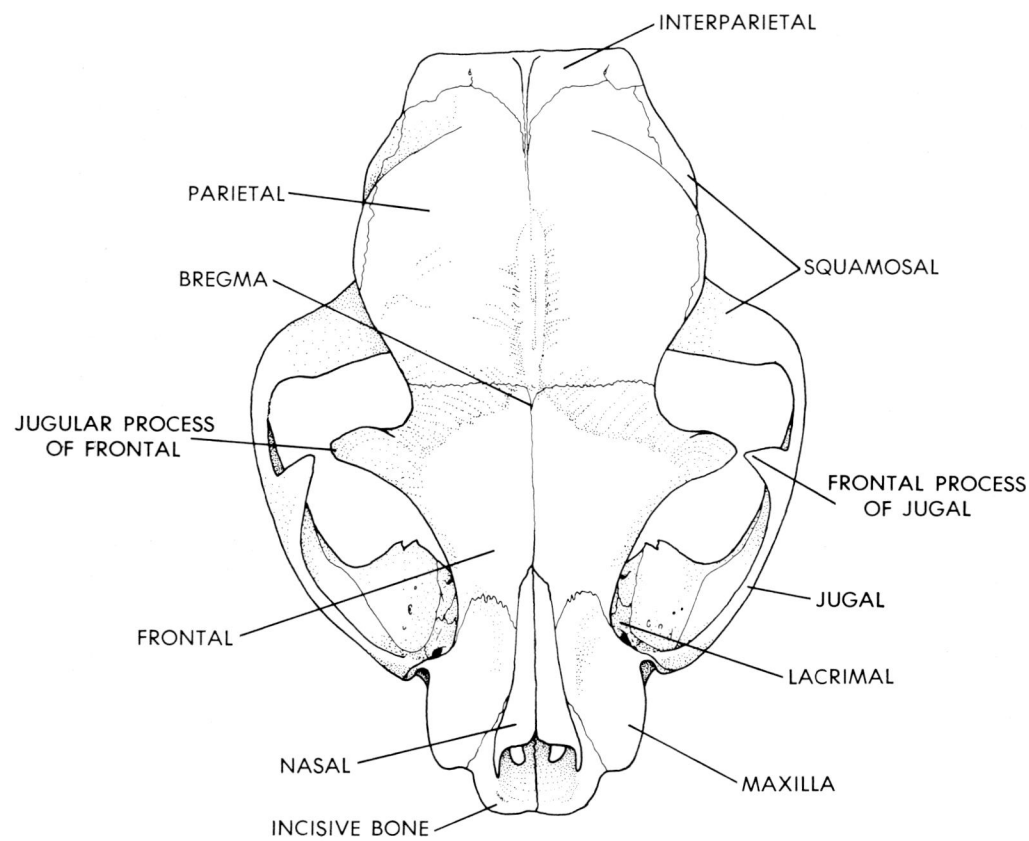

FIGURE 6. Skull, dorsal view.

(articulating with the *basihyal*), *epihyals*, *stylohyals*, and *tympanohyals* (attached to the tympanic bulla). The caudal horn consists of a single pair of *thyrohyals* between the basihyal and the thyroid cartilage of the larynx.

11. Middle Ear Bones (see fig. 70). These three bones, *malleus, incus,* and *stapes,* are described with the ear, p. 100.

C. Dentition. The teeth of mammals occur in two sets: a *deciduous* set of teeth occurs in the young, subadult, and is replaced by a *permanent* dentition in the adult. The ages at which the deciduous teeth of the cat are shed and replaced by permanent teeth are indicated in appendix A. The cat has four types of teeth: *incisors, canines, premolars,* and *molars.* Each incisive bone contains three incisors; each maxilla has one canine, three premolars, and one molar; each mandible has three incisors, one canine, two premolars, and one molar. This may be expressed in a formula as follows:

$$I\frac{3}{3},\ C\frac{1}{1},\ P\frac{3}{2},\ M\frac{1}{1} \times 2 = 30.$$

The deciduous dentition has a pattern similar to that of permanent teeth except the molars of the upper and lower dentitions are not present and most of the teeth are smaller than the permanent teeth.

The first, most rostral, premolar teeth are early evolutionary losses, thus the three premolar teeth are the second, third, and fourth premolars.

The canine teeth have a single cusp and the incisor teeth have a single transverse ridge on each tooth, but the premolars and molars have distinctive cusps with specific names.

The carnassial (cutting) teeth are the upper fourth premolar and the lower molar. The individual cusps of P^4 are: cingular cusp (most rostral); paracone (middle, behind cingular cusp); metacone (caudal, behind paracone); protocone (medial to paracone). The lower carnassial has only a rostral paraconid and a caudal protoconid. The absence of a "heel" composed of a pair of caudal cusps on the lower molar is a character that sets the cats apart from the other carnivores.

II. Vertebrae, Ribs, and Sternum (figs. 3, 9, 10).
   A. Vertebral Column.
      1. Cervical Vertebrae. There are seven cervical vertebrae in mammals. The first is called the *atlas;* the second, the *axis* (see fig. 9). Each cervical vertebra, except the seventh, can be recognized by the transverse foramen, a small hole on either side of the vertebral canal for the passage of the vertebral artery.
      2. Thoracic Vertebrae. There are thirteen of these. They may be identified by their long dorsal spine (see fig. 9), and by the fact that ribs are attached to them.

      Nine of the thirteen pairs of ribs are known as *true ribs*, for they are attached to the sternum by *costal cartilages*. The next three pairs of ribs are joined to the ninth rib rather than to the sternum. These are called *false ribs*. The thirteenth pair of ribs does not join the sternum nor the other ribs. They are called *floating ribs*.

      The rib (fig. 3, 9, 10) is composed of a *vertebral portion* (which is ossified or bony), and a *cartilaginous portion*. The vertebral portion is made up of the *head* or *capitulum* (articulating with the centrum of the vertebra), *tuberculum* (articulating with the transverse process), neck (narrow part between the capitulum and the tuberculum), and the *shaft* (longer portion from the tuberculum to the ventral end of the rib). There are neither necks nor tuberculae on the last three pairs of ribs.

      The sternum is composed of eight segments called *sternebrae*. The cranial sternebrum is the *manubrium*, and the caudal one is the *xiphoid process* or the *xiphisternum*.
      3. Lumbar Vertebrae. There are seven of these. They may be readily identified by the large transverse process that projects ventrally and cranially (see fig. 9).
      4. Sacral Vertebrae or Sacrum. This is a single

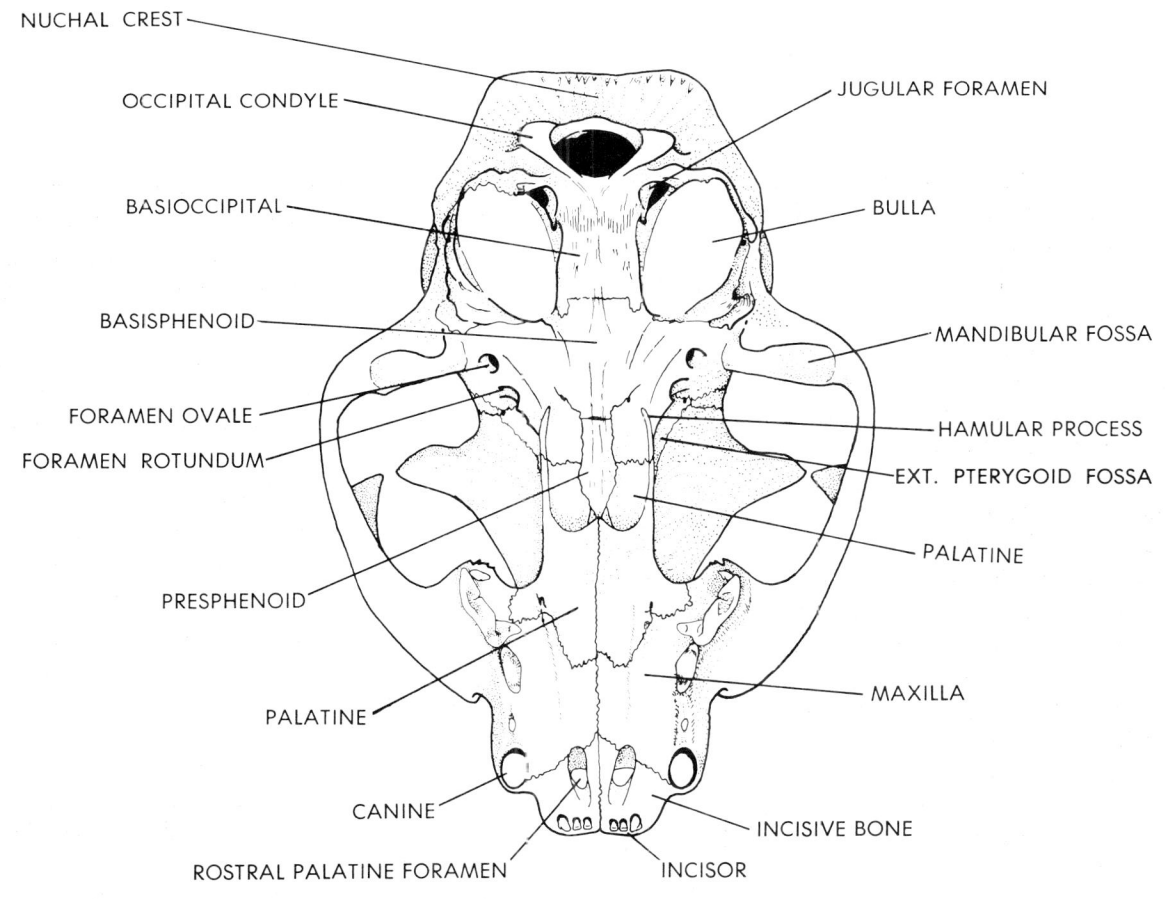

FIGURE 7. Skull, ventral view.

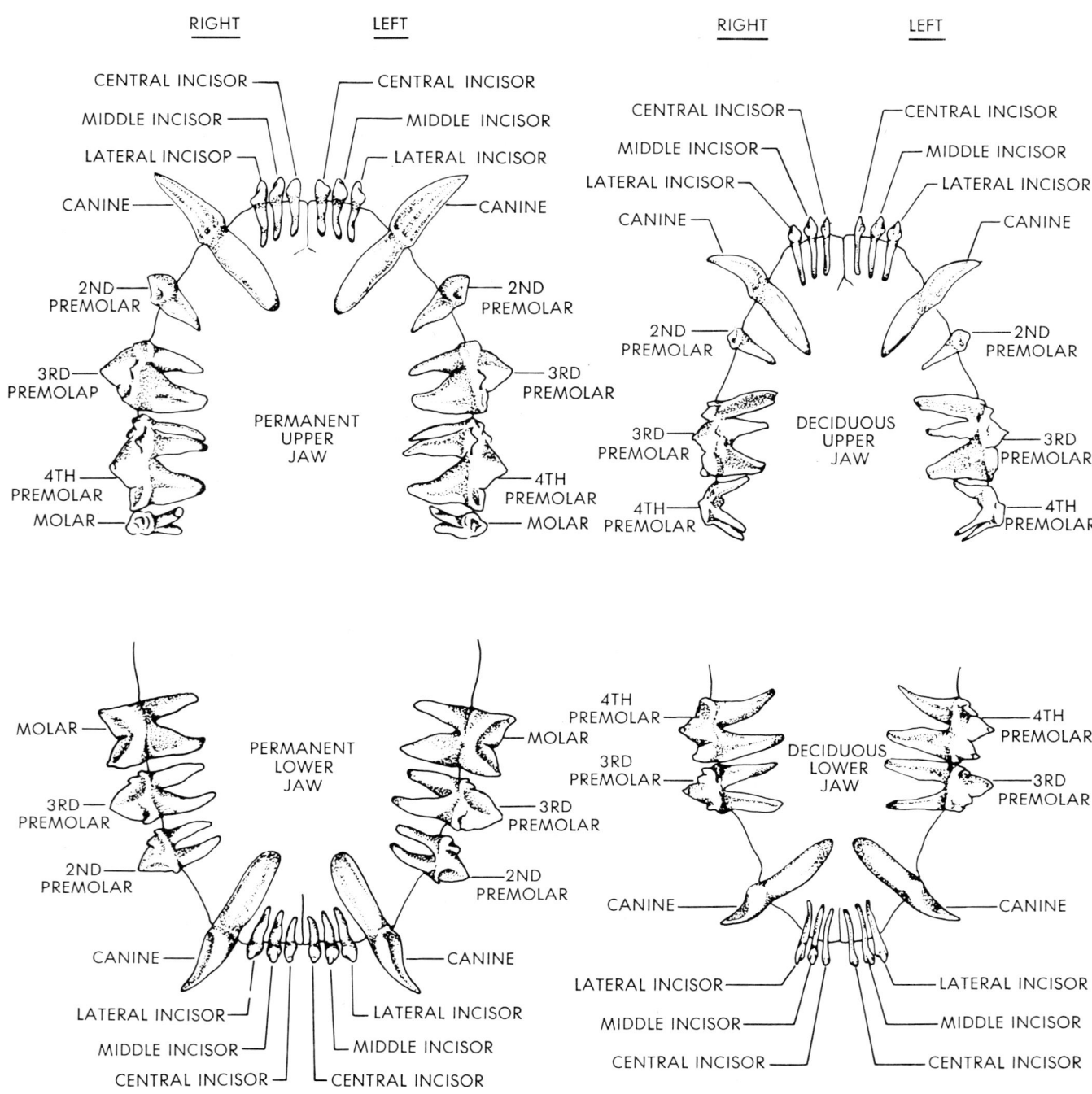

PERMANENT AND DECIDUOUS TEETH OF THE DOMESTIC CAT (Schematic Drawing)

FIGURE 8. Dental patterns of the cat. From Berman, Davis and Stara, 1967. A dental chart of the domestic cat (*Felis catus*, L.). *Laboratory Animal Care*, v. 17, 511-513. By permission.

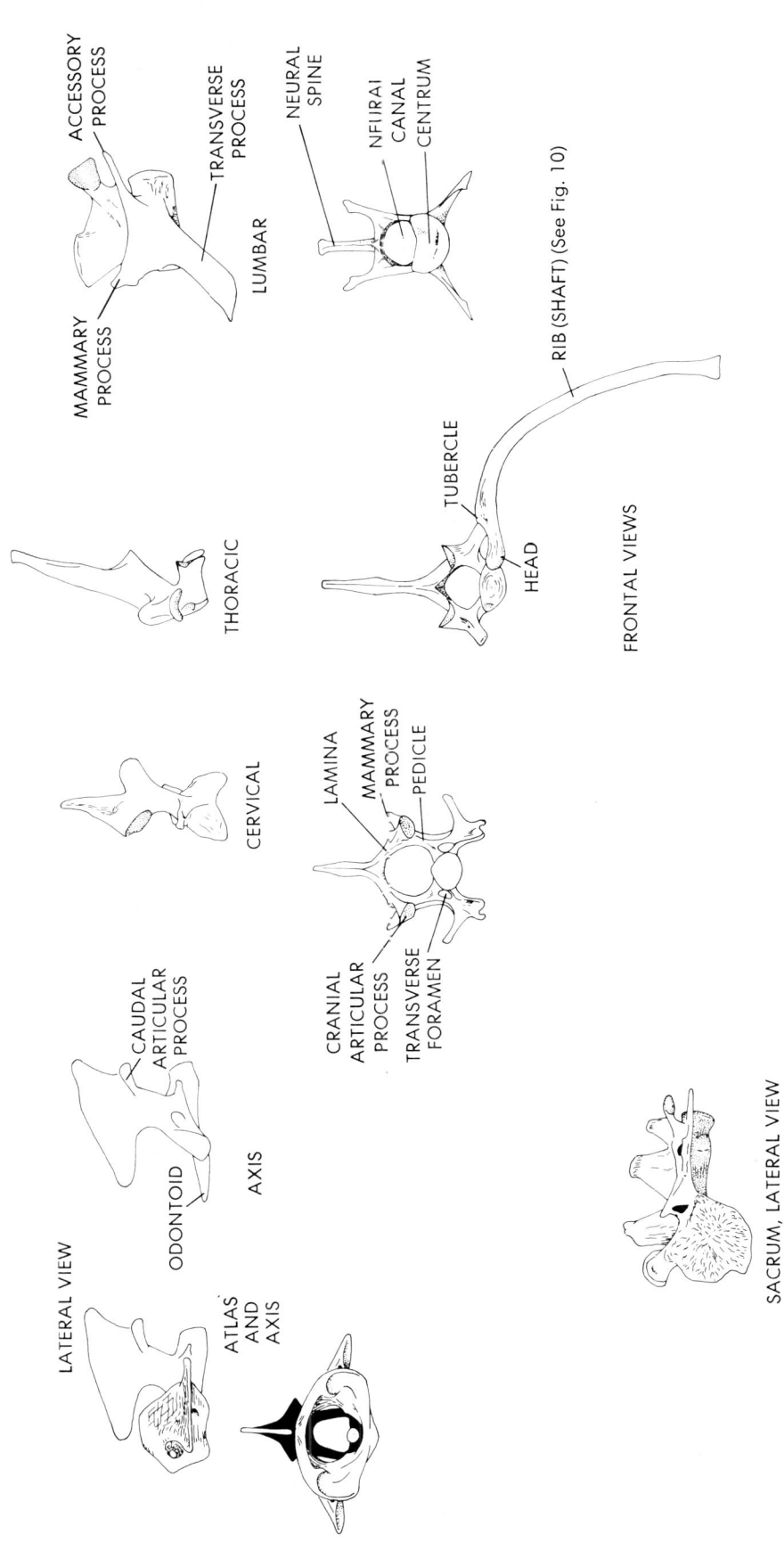

FIGURE 9. Typical vertebrae, lateral (top) and frontal (lower) views.

bone in adults formed by the union of at least three sacral vertebrae.

5. Caudal Vertebrae. The number of these varies a great deal, depending on the length of the tail, but there are usually at least twenty. They may be confused with the bones of the sternum and the phalanges unless one is careful to note the exact shape of each of these three types of bones. Note the "dumbbell" shape of each caudal vertebra, caused by the wide projecting processes at each end of the bone.

6. Typical Vertebra (see fig. 9). On one of the lumbar vertebrae locate the *centrum* (large, ventral body portion), *neural canal* or *vertebral foramen* (through which the spinal cord passes), *neural spine* or *spinous process* (extending dorsally from the neural canal), *transverse processes* (lateral-ventral outgrowths at the base of the neural canal), and the *intervertebral foramina* (narrowed part of the neural arch near the point of attachment of the centrum), where the spinal nerves branch out from the spinal cord.

APPENDICULAR SKELETON

I. Pectoral Girdle and Appendage (figs. 3, 10, 11, 12, 14).

A. Pectoral Girdle.
1. Scapula, paired "shoulder blades." In the cat these bones are the only *functional* part of the pectoral girdle. A *spine* divides the lateral surface into cranial *(supraspinous)* and caudal *(infraspinous) fossa*. A caudal projecting process from the spine is the *metacromion* process, and the ventral continuation of the spine is the *acromion* process. Opposite the acromion process, on the medial surface of the scapula, is the hook-shaped *coracoid* process. Between the acromion and coracoid processes is the *glenoid* fossa for articulation with the head of the humerus.
2. Clavicles are small, paired bones buried in muscles, with no articulation.

B. Pectoral Appendage.
1. Humerus. The proximal end of the humerus is characterized by a *head* that articulates with the scapula and *greater* and *lesser tuberosities* that serve for muscle insertions. The distal end is characterized by a two-planed articular surface: the rounded, lateral half is the capitulum for articulation with the radius, and the concave lateral surface, the *trochlea*, articulates with the ulnar *semilunar notch*.
2. Radius is the lateral bone of the forearm.
3. Ulna is the medial bone of the forearm. The proximal end of the ulna has a *semilunar notch* that articulates with the humerus and a projection beyond the semilunar notch called the *olecranon process*.
4. Carpal bones (fig. 14). The names of these wrist bones are variable, as indicated on page 14.
5. Metacarpals are the bones of the paws. There are five metacarpals in each manus, one for each digit.
6. Phalanges are the bones of the digits. The first digit (pollex) has only two phalanges but the other four digits each have three phalanges. All five digits of each hand have terminal horny claws on the distal phalanges.

II. Pelvic Girdle and Appendage (figs. 3, 13, 15)

A. Pelvic Girdle.
1. Innominate bone is composed of four separate units that fuse together to form each innominate. The two innominates are joined in the ventral midline by the *pubic symphysis* and articulate dorsomedially with the sacrum by the *iliosacral* joint.
   a. Ilium in the cranial-most portion of the innominate.
   b. Ischium is the dorsocaudal portion of the innominate.
   c. Pubis is the ventrocranial portion of the innominate.
   d. Acetabular bone (condyloid bone) is a small triangular bone within the acetabulum.
   e. Acetabulum is the socket for articulation with the head of the femur.
   f. Obturator foramen is a large opening in the caudal half of the innominate, between the ischium, pubis, and acetabulum.
2. Femur is the large bone of the thigh. The proximal end has a head that articulates with the acetabulum and *greater* and *lesser trochanters* for muscle attachments. The head

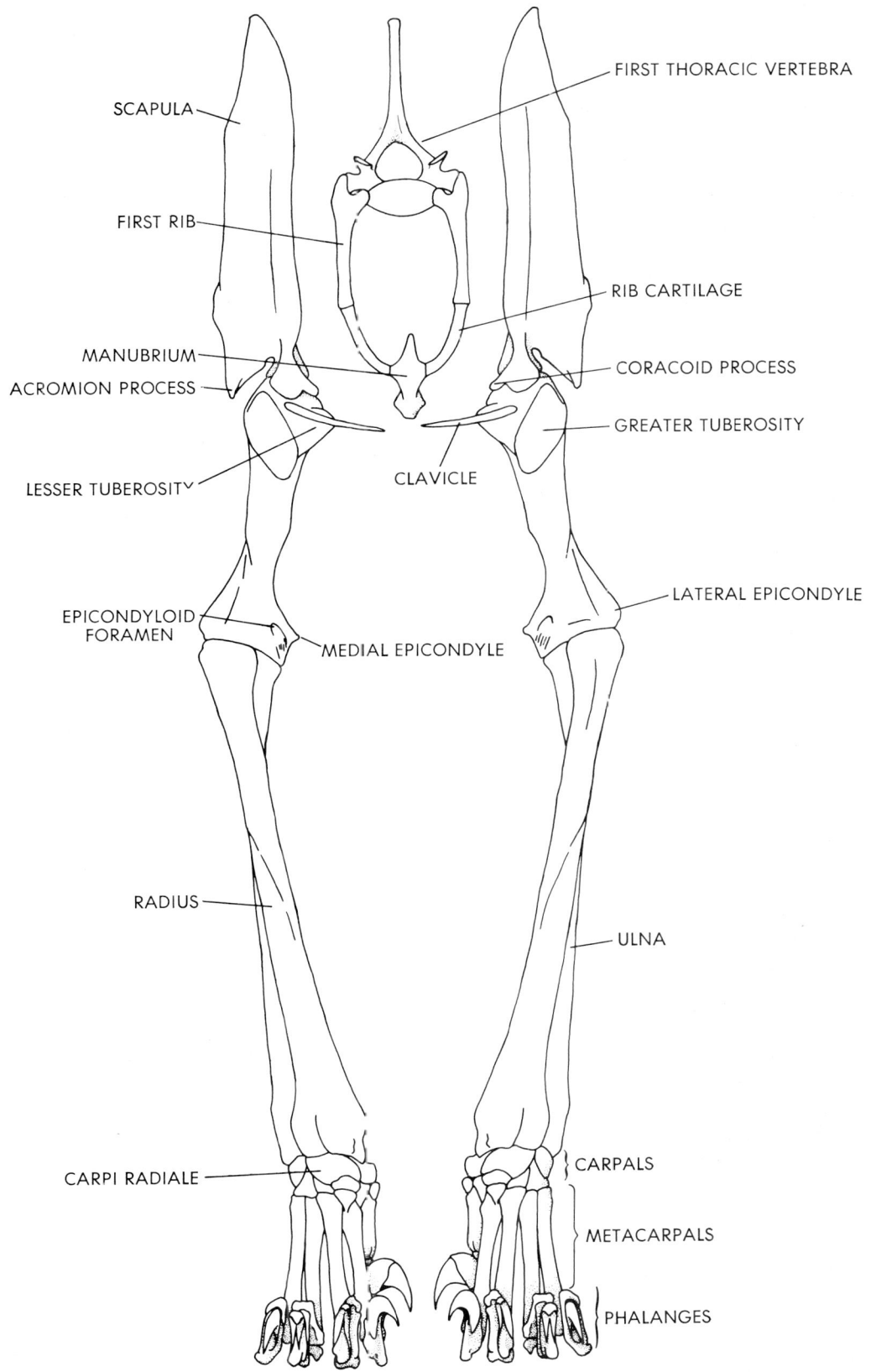

FIGURE 10. Pectoral girdle and limbs, frontal view.

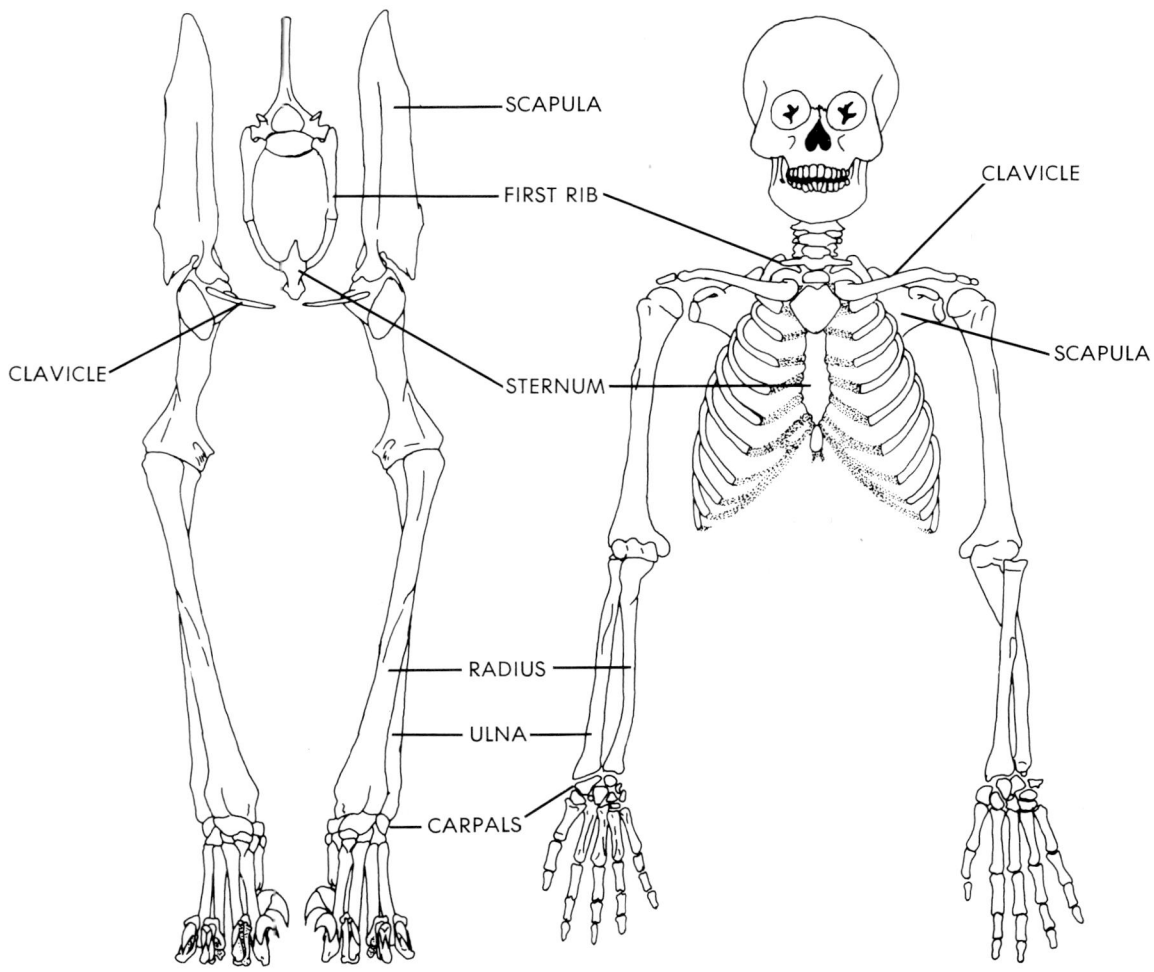

FIGURE 11. Comparison of human and cat girdles and limbs, frontal view.

of the femur is set at an angle of about 45° to the *shaft* of the femur. The distal end has two articular *condyles* separated by an *intercondyloid fossa*.

3. Tibia is the large medial bone of the leg. The proximal end has two articular condyles separated by a spine. The distal end has the *medial malleolus*.

CARPAL BONES

CAT	ALTERNATIVE MAMMALIAN	HUMAN
Radial carpal	Scaphoid	Scaphoid
----------	Intermediate [Lunate]	Lunate
Ulnar carpal	Triquetrum	Triquetrum
Accessory carpal	Pisiform	Pisiform
Carpal I	Trapezium	Trapezium
Carpal II	Trapezoid	Trapezoid
Carpal III	Capitate	Capitate
Carpal IV	Hamate	Hamate
Radial sesamoid	----------	----------

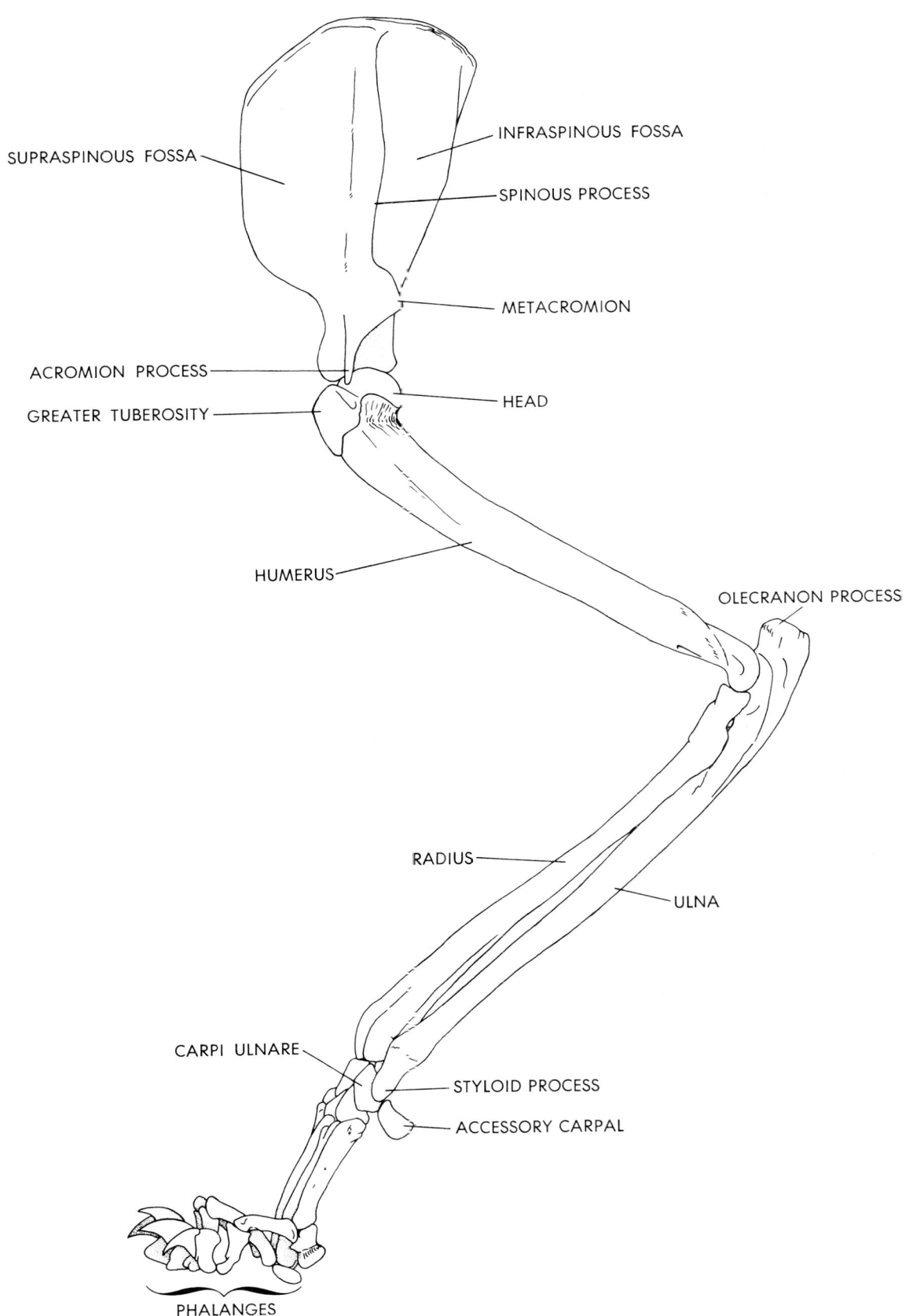

FIGURE 12. Pectoral girdle and limb, lateral view.

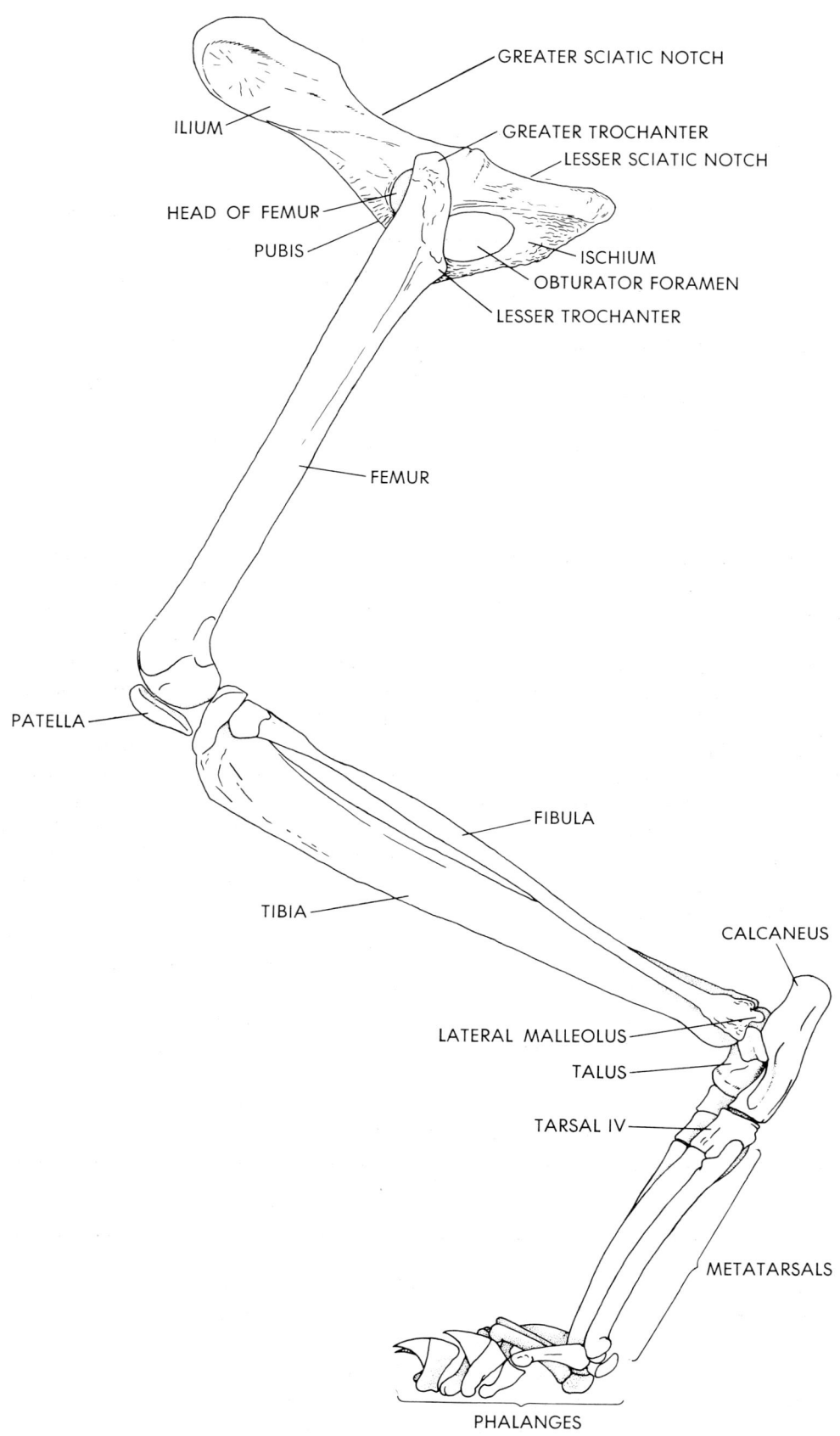

FIGURE 13. Lateral view of pelvic girdle and limb.

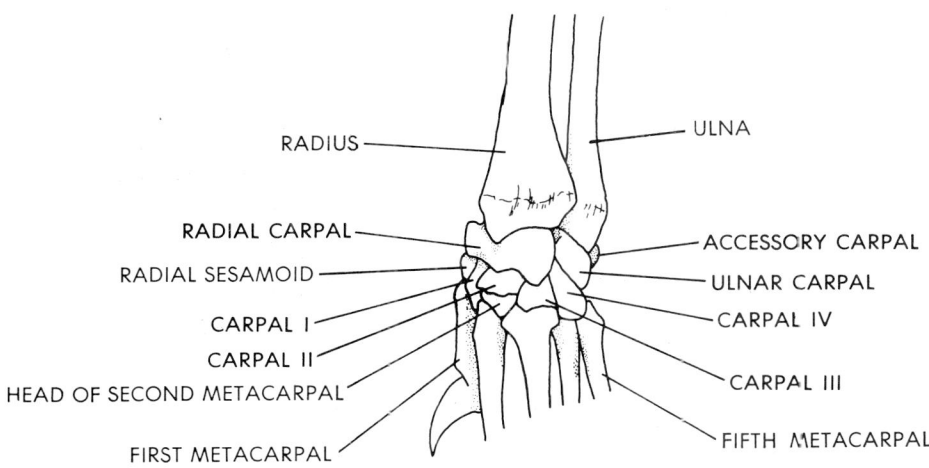

FIGURE 14. Carpal bones, frontal view.

4. Patella is a small, sesamoid bone cranial to knee joint, the "knee cap."
5. Fibula is the slender, splintlike lateral bone of the leg.
6. Tarsals are the ankle bones. The variability in the names of individual tarsal bones will be evident in the chart below.
7. Metatarsals are the bones of the foot or *pes*. The cat has five metatarsals, one for each digit and a rudimentary first metatarsal just distal to the external cuneiform.
8. Phalanges are the bones of the digits. The cat has only four digits on the foot, and there are three phalanges in each digit.

ARTICULATIONS

Joints or articulations are often thought of as the union of two bones. Actually the skeleton originates as a single unit in the embryonic mammal and joints are separations between subunits. Skeletal structures are constructed of two or three tissues; bone, fibrous tissue, and cartilage. The environment of the formative cells (mesenchyme) determines the type of tissue that will form. A good blood supply (vascularization) with compaction of the mesenchyme produces bone; a poor vascularization with compaction produces cartilage. Fibrous tissue is formed by vascularized mesenchyme that is under tension rather than compaction. Consequently, the embryonic skeleton that forms before advanced development of the circulatory system is composed of cartilage. Musculature flexes the cartilage and with the formation of blood vessels the tension applied by contracting muscles produces fibrous tissues in the region of flexion, thus allowing the formation of a joint. The actual shape and construction of the joint surface is dependent on (1) the various muscle attachments that move the parts of the skeleton,

TARSAL BONES

CAT	ALTERNATIVE MAMMALIAN	HUMAN
Talus	Talus	Talus
Calcaneus	Calcaneus [Os calcis]	Calcaneus
Central tarsal	Navicular	Naviculare
Tarsal I	Medial cuneiforme	Medial cuneiforme
Tarsal II	Intermediate cuneiforme	Intermediate cuneiforme
Tarsal III	Lateral cuneiforme	Lateral cuneiforme
Tarsal IV	Cuboideum	Cuboideum

(2) the vasculature to the different areas of the joint, and (3) the forces applied to specific regions of the joint by muscle contraction or body weight.

Those joints that allow considerable movement may be classified in six major anatomical groups; *spheriodal* (ball and socket), *ellipsoidal*, *sellar* (saddle-shaped), *ginglymus* (hinge), *trocoid* (pivot), and *arthroidal* (plane) joints.

These six types of joints may be moved in one (sellar and ginglymus), two (ellipsoidal), or three (spheroidal) directions.

Two excellent examples of joints in the mammal are the hip joint, which has the maximum planes of movement, and the knee, with special ligaments that limit movement to a single plane. Trochoid and arthroidal joints, permitting movement in only one direction, are quite specialized. The trocoid is a pivot joint such as that between the atlas and axis, or between the radius and ulna. The arthroidal joint occurs at the wrist or ankle where the radius and ulna glide over the carpals and the tibia glides over the tarsals.

Sellar and ginglymus joints allow movements in only one plane and therefore allow only flexion and extension of the joint (e.g., between the humerus and ulna in fig 12). Ellipsoidal joints may be flexed and extended as well as abducted and adducted (the carpi ulnare in fig. 12). Spheroidal joints flex and extend, abduct and adduct, and may also circumduct (fig. 15).

A typical diarthroidal (synovial) joint is encapsulated and fluid-filled. The synovial fluid is produced by the synovial membrane and lubricates the articular surfaces of the joint thus reducing friction and wear.

THE HIP JOINT

The hip joint of the cat allows movement in three planes; flexion-extension, abduction-adduction, and rotation or circumduction. This joint and the shoulder are the only joints in the body with three degrees of freedom of movement.

The ligament binding the femur head to the acetabulum (pelvic socket) consists of a capsular ligament surrounding the joint between the cartilage lip of the acetabulum and the neck of the femur. Thickened portions of the capsular ligament are distinguished as separate ligaments; the *iliofemoral* ligament on the cranial border of the capsule, the *ischiofemoral* ligament on the caudal border, and the *pubofemoral* ligament on the ventral border. In addition a ligament, the *transverse acetabular*, extends transversely across the acetabular notch forming the ventromedial wall of the acetabulum.

A ligamentum teres extends from the fovea in the head of the femur to the acetabular fossa located entirely within the capsular ligament. The ligamentum teres is so situated, from a dorsal attachment in the acetabulum to a ventral femoral attachment, that adduction or flexion compresses the ligament against a fat pad in the acetabular fossa. This movement drives the synovial fluid into the dorsal half of the joint cavity, thus lubricating the dorsal head of the femur and the roof of the acetabulum. The dorsal femur head and the acetabular roof are the greatest pressure bearing areas of the joint. Since the fluid normally flows into the ventral half of the joint, the ligamentum teres is essential to pump the fluid to the areas of greatest wear.

The capsular ligament serves primarily to limit the movement of the joint.

THE KNEE JOINT

As the leg moves to a straight position (fig. 17), the wedgelike lateral meniscus is pulled forward by the meniscofemoral ligament and "locks" the joint in the extended position. The popliteal muscle originates on the lateral meniscus and contraction of this muscle pulls the meniscus back so the now "unlocked" joint may be flexed.

The rounded condyles of the femur roll over the relatively flat tibia much like wheels roll over a flat surface (fig. 18). A pair of ligaments cross in the midline of the joint between the condyles of the femur and the menisci of the cranial border of the tibia to the caudal border of the femur (cranial cruciate ligament) and from the caudal border of the tibia to the cranial border of the femur (caudal cruciate ligament). These ligaments act as elastic bands so the caudal cruciate ligament is stretched as the joint is flexed and the cranial cruciate ligament is stretched as the joint is extended (straightened).

The rolling motion of the femur over the tibia causes less wear than would a sliding or pivotal movement, but this arrangement also requires wedgelike menisci and several ligaments to direct (cruciate) and hold (collateral) the joint in place. Removal of the locking meniscus by the contraction of the flexing muscles will allow the elastic cruciate ligament to again direct the movement of the knee.

SUGGESTED READING

Barnett, C. R., Davies, D. V., and MacConaill, M. A. 1961. *Synovial Joints, Their Structure and Mechanics.* Springfield, Ill.: C. C. Thomas Publishers.

Buckland-Wright, J. C. 1978. Bone structure and the patterns of force transmission in the cat skull (Felis Catus). *Jour. Morph.* 155(1):35-62.

Chiasson, R. B. 1960. The preparation of exploded skulls. *Turtox News* 38(11):274-276.

Chiasson, R. B. 1976. Cusp relationships in a carnassial dentition. *Jour. Ariz. Acad. Sci.* 11(1):30-32.

Freeman, M.A.R. and Wyke, B. 1967. The innervation of the knee joint, an anatomical and histological study in the cat. *Jour. Anat.* 101(3):505-532.

———. 1967. The innervation of the ankle joint, an anatomical and histological study in the cat. *Acta Anat.* 68:321-333.

Frewin, J. 1970. The heamapophyses in the caudal vertebrae of the cat, dog and ox. *Zbl. Vet. Med.*, A., 17:565-572.

Gindhart, P. S. 1972. The effect of seasonal variation on long bone growth. *Human Biology* 44(3):335-350.

Jayne, H. 1898. *Mammalian Anatomy, Part I. The Skeleton of the Cat*. Philadelphia: J. B. Lippincott Co.

Lindsay, F.E.F. 1968. Skeletal abnormalities of a cat thorax. *Brit. Vet. Jour.* 124:306-307.

Parsons, T. S. and Stein, J. M. 1956. A cat skeleton with an anomalous third hind leg and abnormal vertebrae. *Bull. Mus. Comp. Zoology Harvard*, 114(6):315-317.

Riach, I.C.F. 1967. Ossification of the sternum as a means of assessing skeletal age. *Jour. Clin. Path.* 20:589-590.

Smith, R. N. 1969. Fusion of ossification centers in the cat. *Jour. Small Anim. Pract.* 10:523-530.

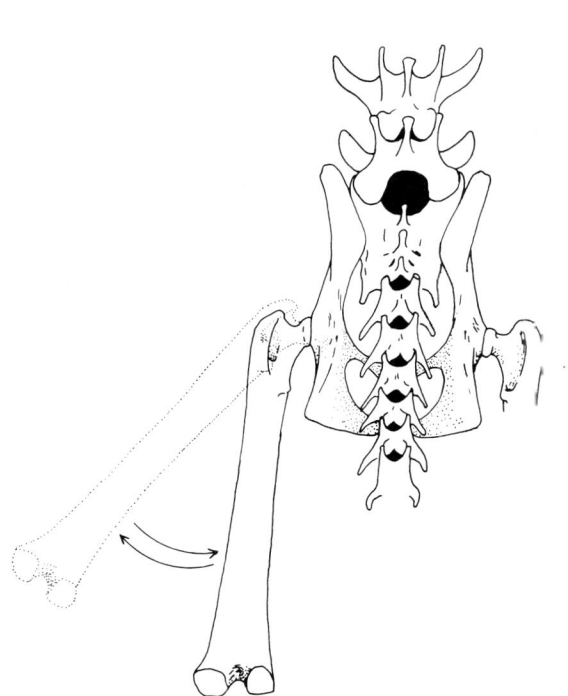

FIGURE 15. Dorsal view of the pelvic vertebrae, girdle, and femur illustrating abduction-adduction (arrows). See fig. 16 for rotation or circumduction of the hip joint.

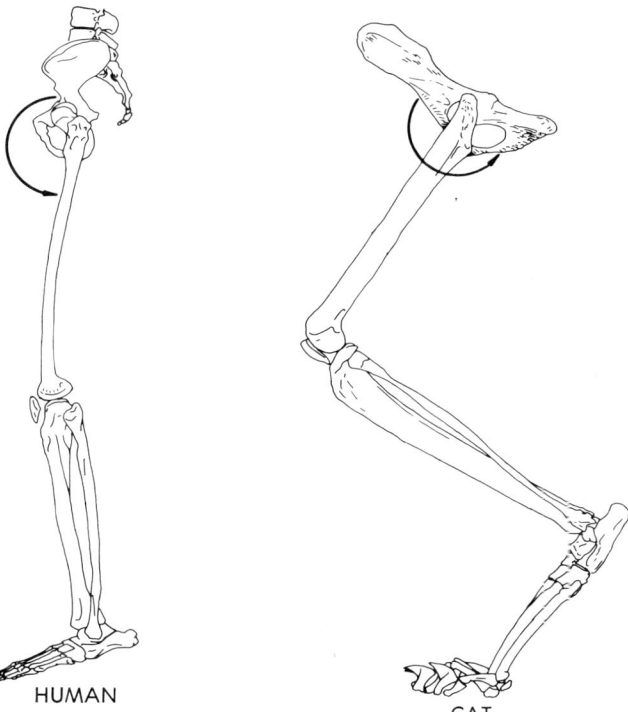

FIGURE 16. Lateral views of the pelvic limb and girdle of human and cat to illustrate thigh movements. In the normal erect position the human limb is fully extended. The cat femur is held at an angle of approximately 90° to the long axis of the body. The arched arrow indicates the movement of the femur at its joint with the acetabulum.

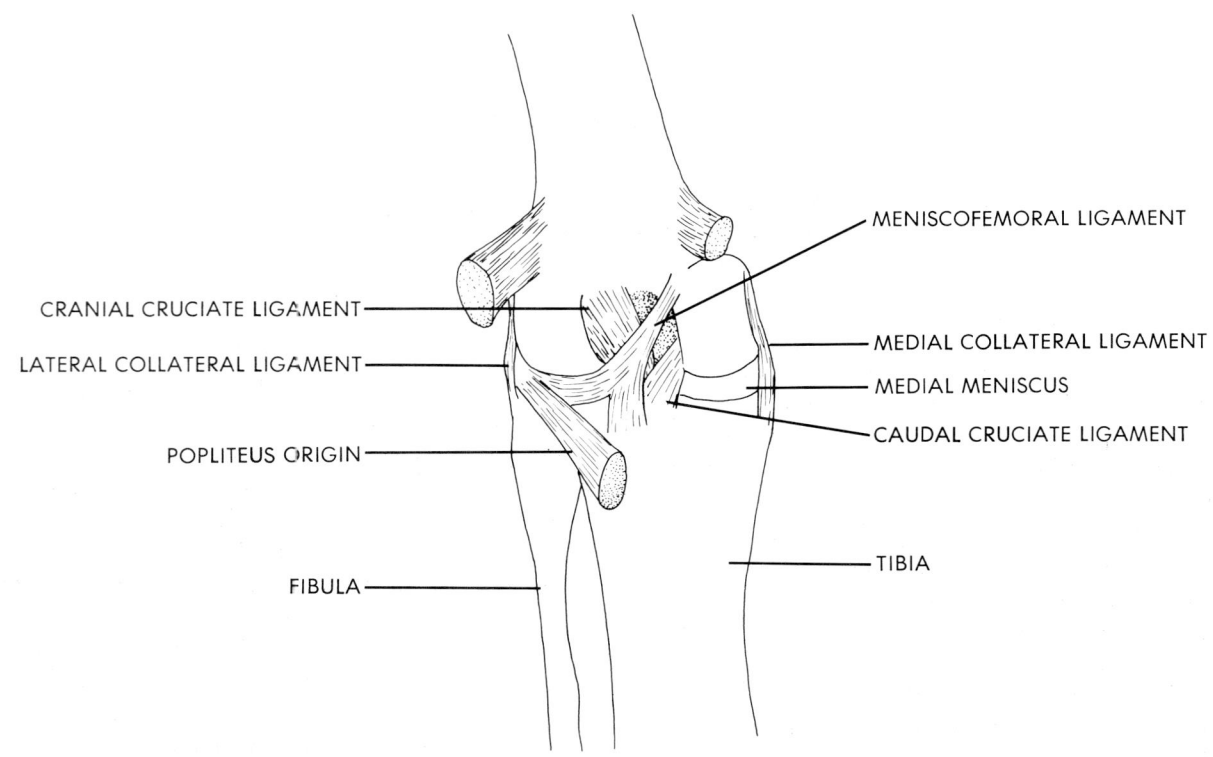

FIGURE 17. Caudal view of the cat knee joint. Much of the capsular ligament is removed to reveal deeper structures.

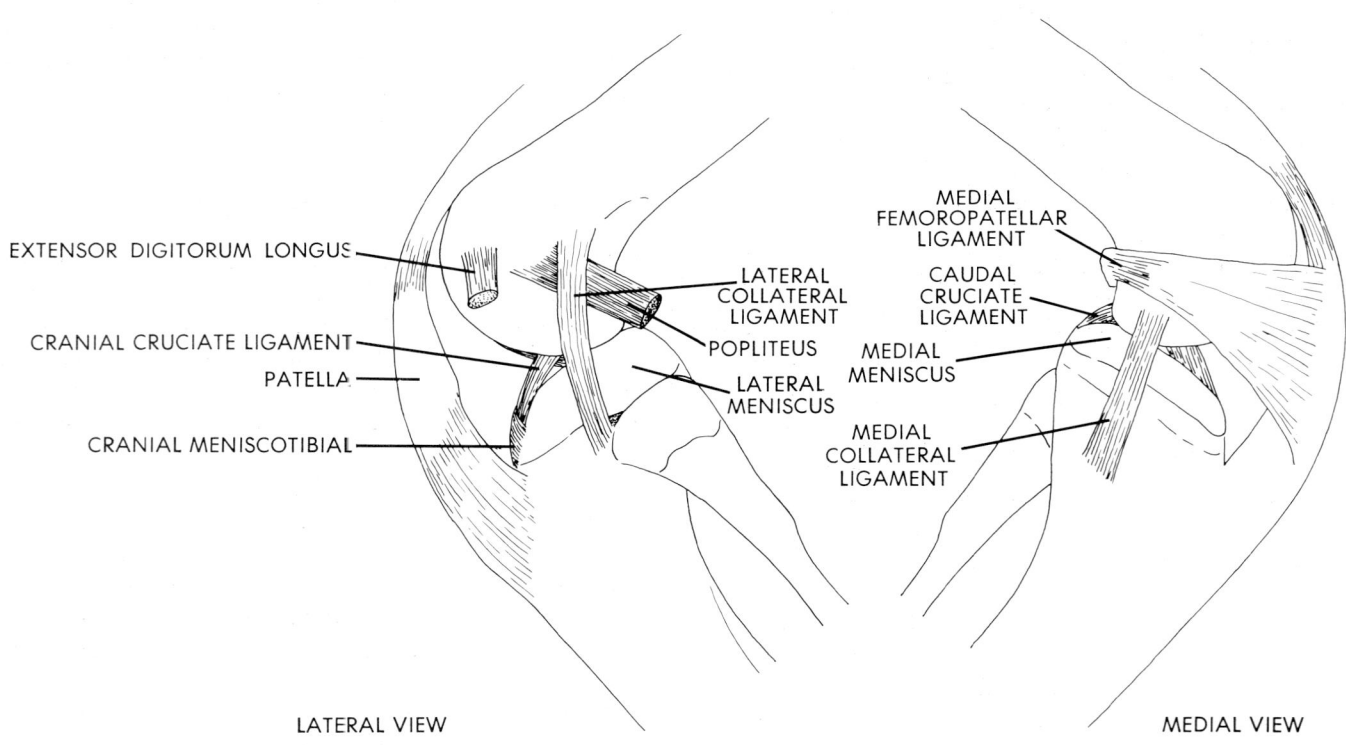

FIGURE 18. Lateral and medial view of the cat knee joint. Much of the capsular ligament is removed. See text.

Chapter 3
Muscular System

SKINNING

Use scissors with one blunt tip. Insert the blunt tip between the skin and the musculature in the opening in the neck made by the preparator when embalming and injecting the blood vessels. Continue the cut caudally in the ventral midline to a point between the two hind limbs and cranial to the genitalia. Carefully separate the skin from the underlying tissues (muscle, fat, and/or mammary gland) with a blunt probe. Again, with your scissors cut from the midline slit on the ventral side to the wrist and make another cut to the ankle on one side of the cat. Next, cut around the wrist and ankle and carefully separate the skin from the underlying tissues.

Special care will be necessary, if the specimen is a female, to clean the mammary glands. Each nipple should be carefully cut on the inside of the skin, leaving as much of the teat with the gland as possible.

Now continue the incision on the ventral side of the neck forward to the chin and ventral lip. Always direct the blunt point of the scissors inward toward the musculature and keep the sharp point outward.

Separate the skin from the underlying tissues by inserting the closed scissors' blades between the skin and cheek muscles and between the eye and upper lip. Then open the scissors' blades. Repeat this procedure with the scissors between the eye and the ear. With a sharp scalpel carefully cut the skin around the lips and the eye while pulling the ventral edge of the skin of the head dorsally. Probe with your fingers around the ear, then cut the ear loose with your scissors.

The *cutaneous maximus* (see p. 22) is a large flank muscle originating in the ventral midline and armpit and inserting in the skin of the flank region. If this muscle is to be left intact on the specimen, it will be necessary to scrape it loose from the skin with a scalpel. To do this, grip the cut edge of the skin in one hand and pull it away from the body. This exposes white connective tissue between the skin and the tan colored muscle. Carefully cut these fibers with your scalpel, pointing the tip of the scalpel toward the skin and avoid cutting the muscle fibers as long as possible. The *latissimus dorsi* insertion is beneath the axillary origin of the *cutaneous maximus*, and the two muscles will be difficult to separate from each other. The mammary gland of the female lies superficial to the ventral origin of the *cutaneous maximus*.

There should be considerable fat in the neck, axillary, and inguinal areas that must be cleared away very carefully. Pick away the fat in a manner that does not damage the blood vessels in the centers of these masses. If your specimen is a lactating female, the mammary glands will be served by branches of the *lateral thoracic* artery and vein cranially and the *superficial epigastric* branches of the *external pudendal* artery and vein. It will be necessary to cut these vessels and to remove the mammary glands on one side of the body in order to see the abdominal musculature.

When cleaning fat and connective tissue from blood vessels and nerves, use a forceps (not a rat-tooth) and pull the strands of tissue distally. Be sure to clean and separate each muscle carefully before cutting it. Cut muscles with scissors perpendicular to the fiber direction and as near the middle (half way between origin and insertion) as possible.

TYPICAL MUSCLE

Muscles are covered with a tough connective tissue known as *fascia*. The outer layer of fascia is spongy (superficial fascia); the inner layer, covering the muscles (deep fascia), is rather thin. Study the muscle on the calf of the leg to become familiar with muscle terms. This is the *gastrocnemius* muscle. The muscle may now be separated into medial and lateral parts (called *heads*). Look carefully for bundles of muscle fibers (*fasciculi*), and for the coverings of each fasciculus (*perimysium*). The thick part of the muscle is the belly; the cords of connective tissue at the ends of the muscle are known as *tendons*. A flat sheet of fascia having the function of a tendon is called an *aponeurosis*.

Each skeletal muscle is connected by connective tissue bands to the bone that it moves. This is the *insertion* of the muscle. The proximal attachment of the muscle (usually attached to a fixed structure) is the *origin*.

The function of the muscle is its *action*, or the work that it does in the body. The origin, insertion, and action of the more common muscles are listed below in tabular form. Directions for cutting specific muscles during your study are given at the appropriate place in the muscle chart and/or on the illustrations.

SKIN MUSCLES

Skin muscles are striated muscles that often lack skeletal attachments. One of these, the *cutaneous maximus*, is a large sheet of skin muscle on the side of the body. This large sheet originates on each side of the fascia of the *latissimus dorsi* and of the ventral pectoral muscles. Fibers of the *cutaneous maximus* are often confused with those of the latissimus dorsi or *xiphihumeralis*.

The other skin muscles are located in the skin of the head and neck. The following list includes the skin muscles found in the cat.

1. Cutaneous maximus
2. Platysma
3. Intermedius scutulorum
4. Corrugator supercilii medialis
5. Orbicularis oculi
6. Corrugator supercilii lateralis
7. Frontoauricularis
8. Levator auris longus
9. Auricularis superior
10. Abductor auris longus
11. Abductor auris brevis
12. Epicranius (occipitalis/frontalis)
13. Zygomaticus (major/minor)
14. Submentalis
15. Auricularis externus
16. Transversus auriculi
17. Rotator auris
18. Caninus
19. Tragicus medialis
20. Tragicus lateralis
21. Helicus
22. Depressor conchae
23. Frontoscutularis
24. Abductor auris inferior
25. Abductor auris superior
26. Abductor auris medius
27. Orbicularis oris
28. Quadratus labii superioris
    a. Levator labii superioris alaeque nasi
    b. Levator labii superioris proprius
29. Levator anguli oris
30. Buccinator
31. Myrtiformis
32. Moustachier
33. Quadratus labii inferioris
34. Transversus menti

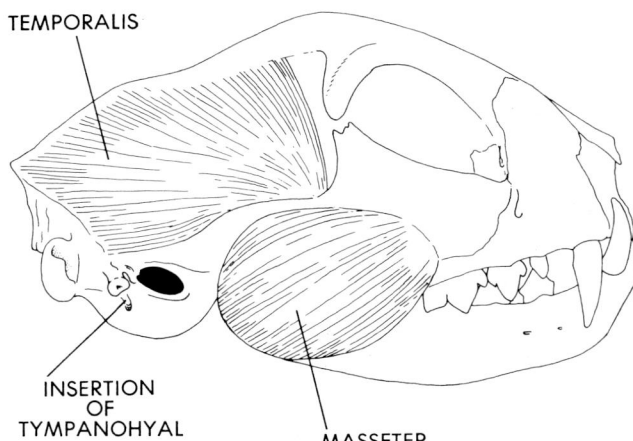

FIGURE 19. Lateral view of jaw muscles.

HEAD

Remove the rostral head of the *Digastricus* on one side and leave the opposite muscle intact. Insert a blunt probe under the lateral constricted portion of the *Digastricus* between the rostral and caudal bellies and then carefully scrape the insertion of the rostral head loose from the mandible.

With a blunt probe scrape the insertion and origin of the *Masseter* loose from the mandible and zygomatic arch and remove the muscle. With bone cutters cut through the two ends of the exposed zygomatic arch and remove the arch. Next place the blunt probe tip on the inner margin of the mandible from which you have removed the *Digastricus* insertion and clear the tissues from the bone. Next cut the mandible with bone cutters at the diastema between the canine and premolars and between the molar and coronoid process and remove this segment. The *Pterygoideus lateralis* and *medialis* muscles are holding the remaining portion of the mandible to the cranium. Inspect these muscles on the medial surface of the mandible.

NAME	ORIGIN	INSERTION	ACTION
Masseter	Zygomatic arch.	Coronoid fossa of mandible.	Elevates mandible.
Temporalis	Temporal fossa of parietals, squamosal, and frontals.	Coronoid fossa of mandible.	Elevates mandible.
Pterygoideus lateralis	External pterygoid fossa of palatine.	Medial surface of mandible.	Elevates mandible.
Pterygoideus medialis	Internal pterygoid fossa of basisphenoid (hamulus).	Medial surface of mandible.	Elevates mandible.
Digastricus	Jugular and mastoid processes of occipital.	Mandible.	Depresses mandible.

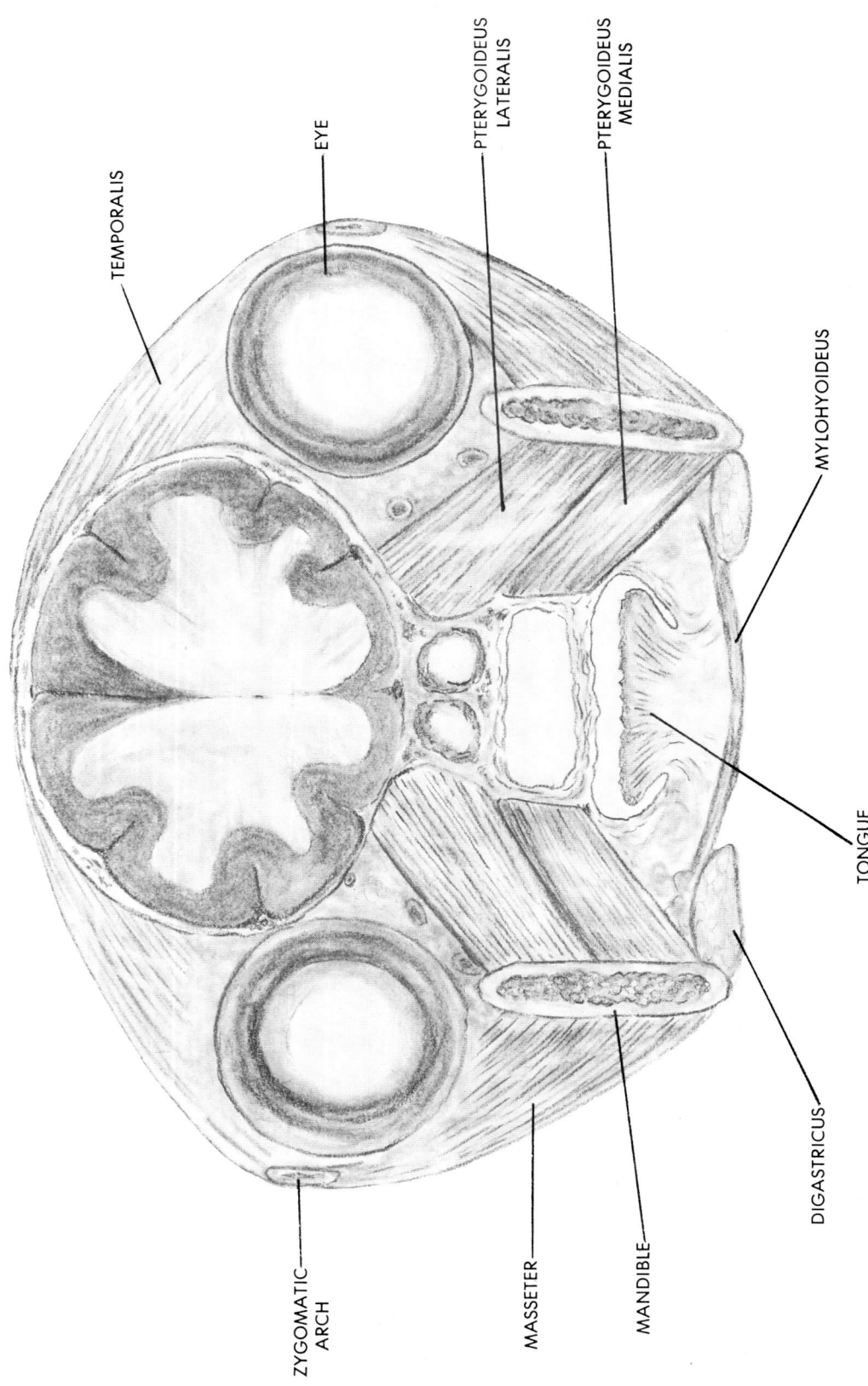

FIGURE 20. Diagrammatic section through the head of the cat to illustrate the arrangement of the pterygoid and other muscles of mastication. In an actual section the origin and insertion of the pterygoids would not be seen at the same level.

HYOID AND TONGUE

The *Mylohyoideus* and *Stylohyoideus* muscles were exposed by dissection of the *Digastricus* (see page 23). Insert a blunt probe beneath the *Mylohyoideus* rostral to the basihyal and separate this muscle from the deeper muscles. Cut the *Mylohyoideus* near the midline with a scalpel against the blunt probe beneath the muscle. Pull the *Mylohyoideus* toward the mandible (where the *Digastricus* was inserted) thus exposing the deeper muscles. The *Geniohyoideus* is near the midline superficial to the *Hyoglossus*. The *Styloglossus* is lateral (near the mandible) to the hyoid muscles. The study of tongue muscles (*Genioglossus, Styloglossus,* and intrinsic muscles) should be postponed until dissection of the oral cavity (Digestive System, Chapter 5).

Transect the *Stylohyoideus* and the sheet of pharyngeal muscle originating from the area just beneath the origin of the *Stylohyoideus*. This will expose the *Ceratohyoideus*.

The *Sternohyoideus, Sternothyroideus,* and *Thyrohyoideus* muscles will be seen with dissection of the neck muscles.

NAME	ORIGIN	INSERTION	ACTION
Stylohyoideus	Stylohyoid bone.	Body of hyoid.	Draw hyoid forward (elevates base of tongue).
Geniohyoideus	Mandible near symphysis.	Body of hyoid.	Draw hyoid (and tongue) forward.
Occipitohyoideus	Jugular process of occipital.	Stylohyoid bone.	Draw hyoid (and tongue) back.
Mylohyoideus	Body of mandible.	Midline of throat to opposite muscle.	Raises floor of mouth.
Ceratohyoideus	Ceratohyoid and epihyoid.	Thyrohyoid.	Draws horns of hyoid together.

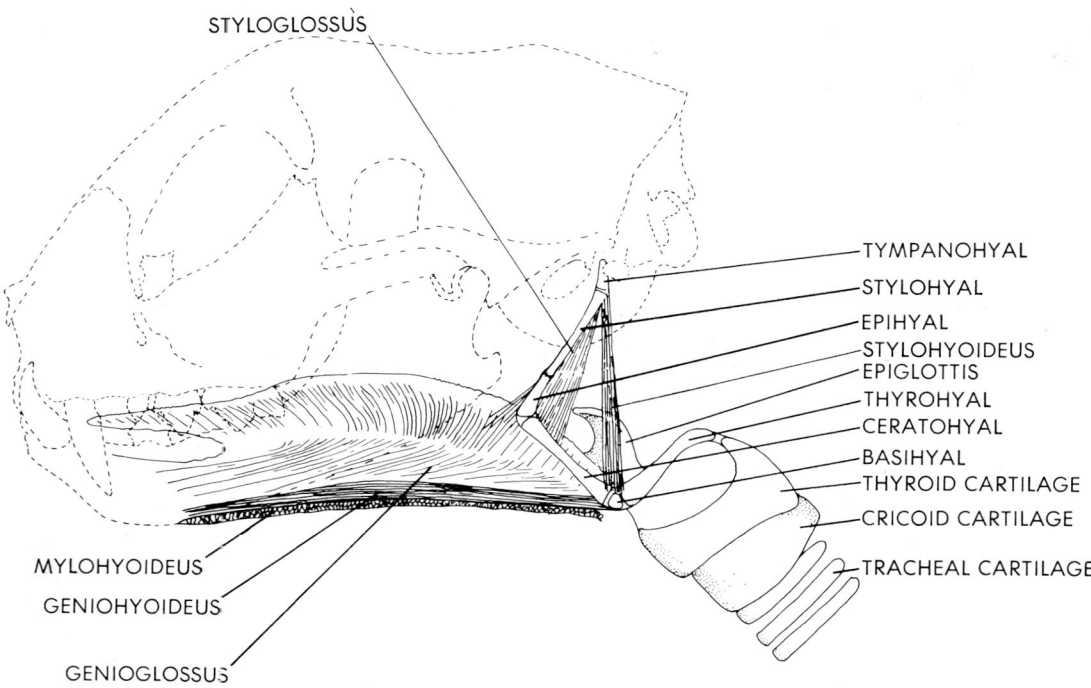

FIGURE 21. Lateral view of the hyoid, tongue, larynx and associated musculature.

NECK

The external jugular vein and its anastomosing link should be cut and reflected in order to dissect the neck muscles. Separate the *Sternomastoideus* from deeper muscles with a blunt probe and transect it half way between its origin and insertion on the same side as the dissected *Digastricus* and *Mylohyoideus* muscles. Separate the *Sternohyoideus* from the *Sternothyroideus* and transect the *Sternohyoideus*. The *Thyrohyoideus* will be seen beneath the rostral end of the *Sternohyoideus*.

NAME	ORIGIN	INSERTION	ACTION
Sternomastoideus (Sternocephalicus)	Manubrium of sternum and fascia of transverse pectoralis.	1. Cranial portion on mastoid. 2. Caudal portion on Lambdoidal ridge of occipital.	Turns head and depresses snout.
Cleidomastoideus	Mastoid process of temporal bone.	Clavicle (medial to cleidocephalicus).	Turns head and depresses snout or draws clavicle forward.
Sternohyoideus	First rib cartilage.	Basihyoid	Draws hyoid caudally.
Sternothyroideus	First rib cartilage beneath sternohyoideus.	Thyroid cartilage of larynx.	Draws larynx caudally.

The deeper neck muscles will be dissected after the dissection of the BACK AND TRUNK muscles.

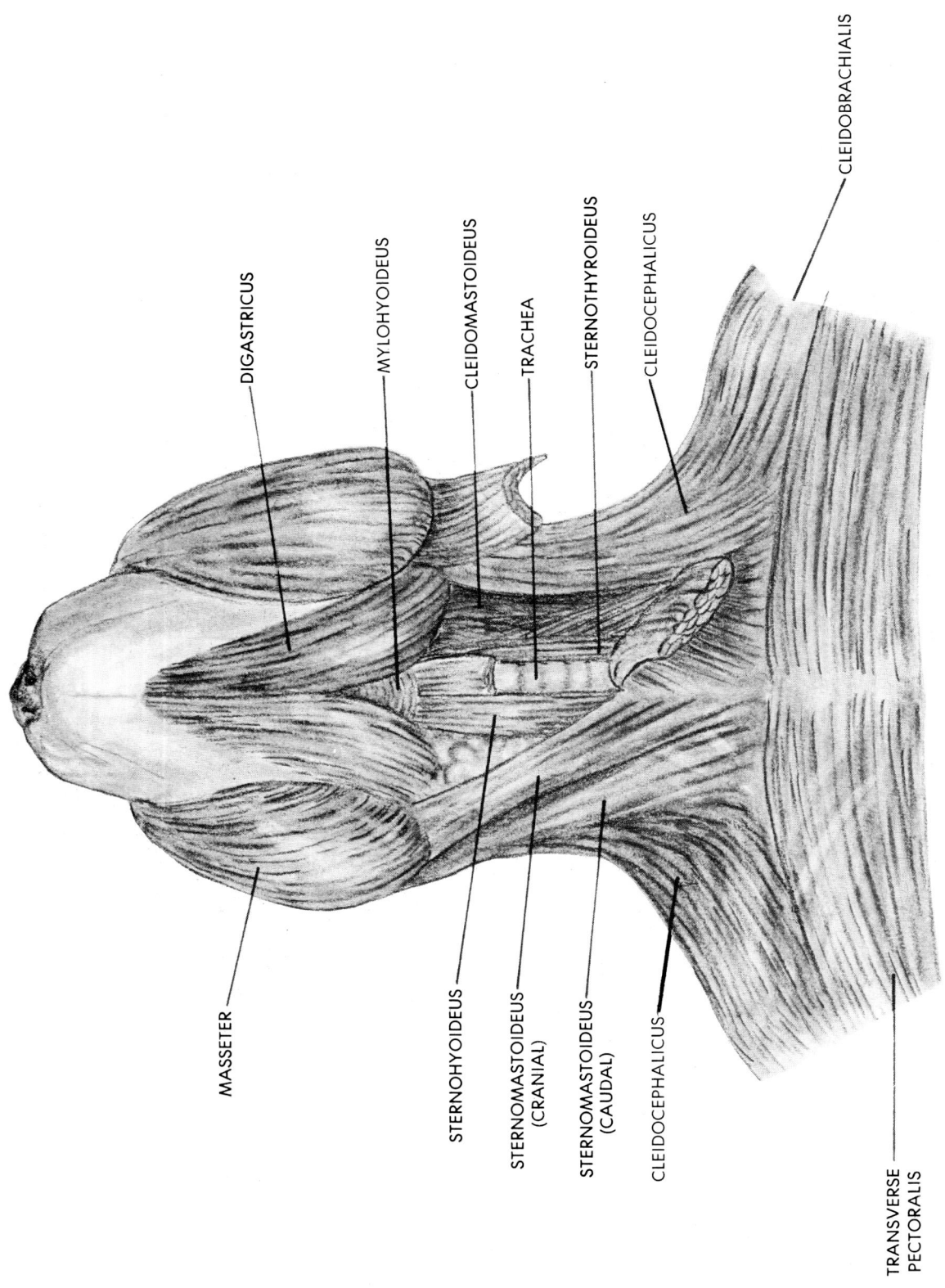

FIGURE 22. Ventral view of the throat and neck muscles of the cat.

SHOULDER

Identify the following muscles: *Cleidocephalicus*, *Cleidobrachialis*, *Trapezius*, and *Latissimus dorsi* from the *Trapezius* and *Pectoralis*. Note the *Cutaneous maximus* (p. 22) inserts on the *Latissimus dorsi* and may be confused with it.

As before, separate the muscles with a blunt probe before transecting. Note the cephalic vein crosses over the *Cleidocervical* portion of the *Brachiocervicalis* muscle at the shoulder. Now cut the *Cleidocervicalis* carefully. A branch of the spinal accessory nerve passes caudally on the neck just beneath the *Cleidocervicalis* and *Trapezius* muscles. Do not cut this nerve. Next transect the trapezius. Cut the *cleidobrachialis* to expose the *Deltoideus*. Next loosen and then cut the scapular and acromial portions of the *Deltoideus*. Flex the brachium against the scapula and insert a blunt probe between the scapula and *Teres major* and loosen the *Teres major* then transect it. The *Teres minor* may be seen just distal to the *Teres major* and lateral to the insertion of the long head of the *Triceps*.

It is necessary to dissect the breast muscles before completing the dissection of the shoulder muscles on the medial side of the shoulder.

SHOULDER (Figures 23, 24, 25, 26, 30, 31)

NAME	ORIGIN	INSERTION	ACTION
Deltoideus			
pars scapularis	Spine of scapula.	Deltoid ridge of humerus.	Flexes humerus.
pars acromialis	Acromion process of scapula.	Mainly on the Pars Scapularis.	Flexes humerus.
pars clavicularis (cleidobrachialis)	Clavicle and cleidocephalicus.	Ulna, distal to semilunar notch.	Flexes antibrachium.
Supraspinatus	Supraspinous fossa of scapula.	Greater tuberosity of humerus.	Extends humerus.
Infraspinatus	Infraspinous fossa of scapula.	Greater tuberosity of humerus.	Rotates humerus outward.
Teres minor	Glenoid border of scapula.	Deltoid tuberosity of humerus.	Rotates humerus outward.
Subscapularis	Subscapular fossa of scapula.	Lesser tuberosity of humerus.	Adducts humerus.
Teres major	Glenoid border of scapula near vertebral border.	Fuses with latissimus dorsi and inserts on humerus.	Flexes humerus.
Coracobrachialis	Coracoid process.	Medial surface of humerus.	Adducts humerus.
Serratus ventralis thoracis	First 10 ribs.	Vertebral border of scapula.	Depresses scapula. (See fig. 26.)
Serratus ventralis cervicis	Last 5 cervical vertebrae.	Vertebral border of scapula, with above.	With above, depresses scapula. (See fig. 26.)
Trapezius			
pars thoracica	Spinous processes of thoracic vertebrae.	Fascia covering supra- and infraspinatus and spine of scapula	Draws scapula caudally and medially.

NAME	ORIGIN	INSERTION	ACTION
pars cervicalis	Spinous processes of axis to 3rd thoracic vertebrae.	Metacromion and spine to scapula.	Draws scapula medially.
Brachiocephalicus cleidocephalicus	Lambdoidal crest of occipital.	Clavicle and cleidobrachialis muscle.	Draws clavicle forward and helps flex antibrachium.
cleidobrachialis	See *Deltoideus-pars clavicularis*—origin, insertion, and action		
Rhomboideus capitis	Lambdoidal ridge of occipital.	Vertebral border of scapula cranial to *Rhomboideus thoracis*.	Draws scapula forward.
Rhomboideus cervicis *et thoracis*	Supraspinous ligament of cervical vertebrae and first 4 thoracic vertebrae.	Vertebral border of scapula.	Draws scapula medially and dorsally.
Omotransversarius	Transverse process of atlas and basioccipital.	Metacromion of scapula.	Draws scapula forward.
Latissimus dorsi	Neural spines of 4th thoracic to 6th lumbar vertebra.	With teres major on humerus.	Flexes humerus.

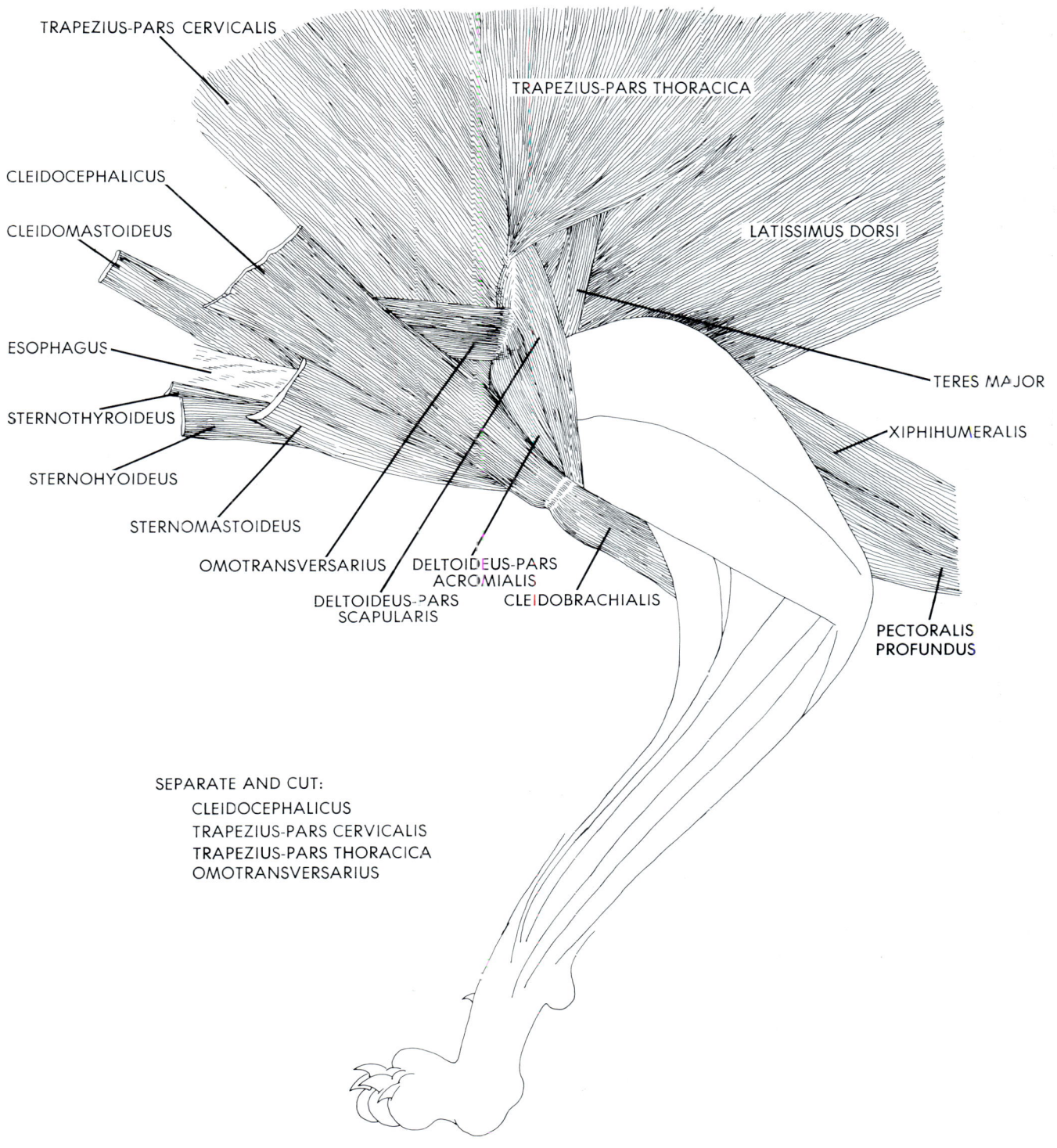

FIGURE 23. Superficial lateral muscles of the shoulder, brachium, and arm. Carefully separate, clean, and section the following muscles to expose the muscles illustrated in fig. 24: *Trapezius-pars cervicalis, trapezius-pars thoracica, omotransversarius,* and *cleidocephalicus.*

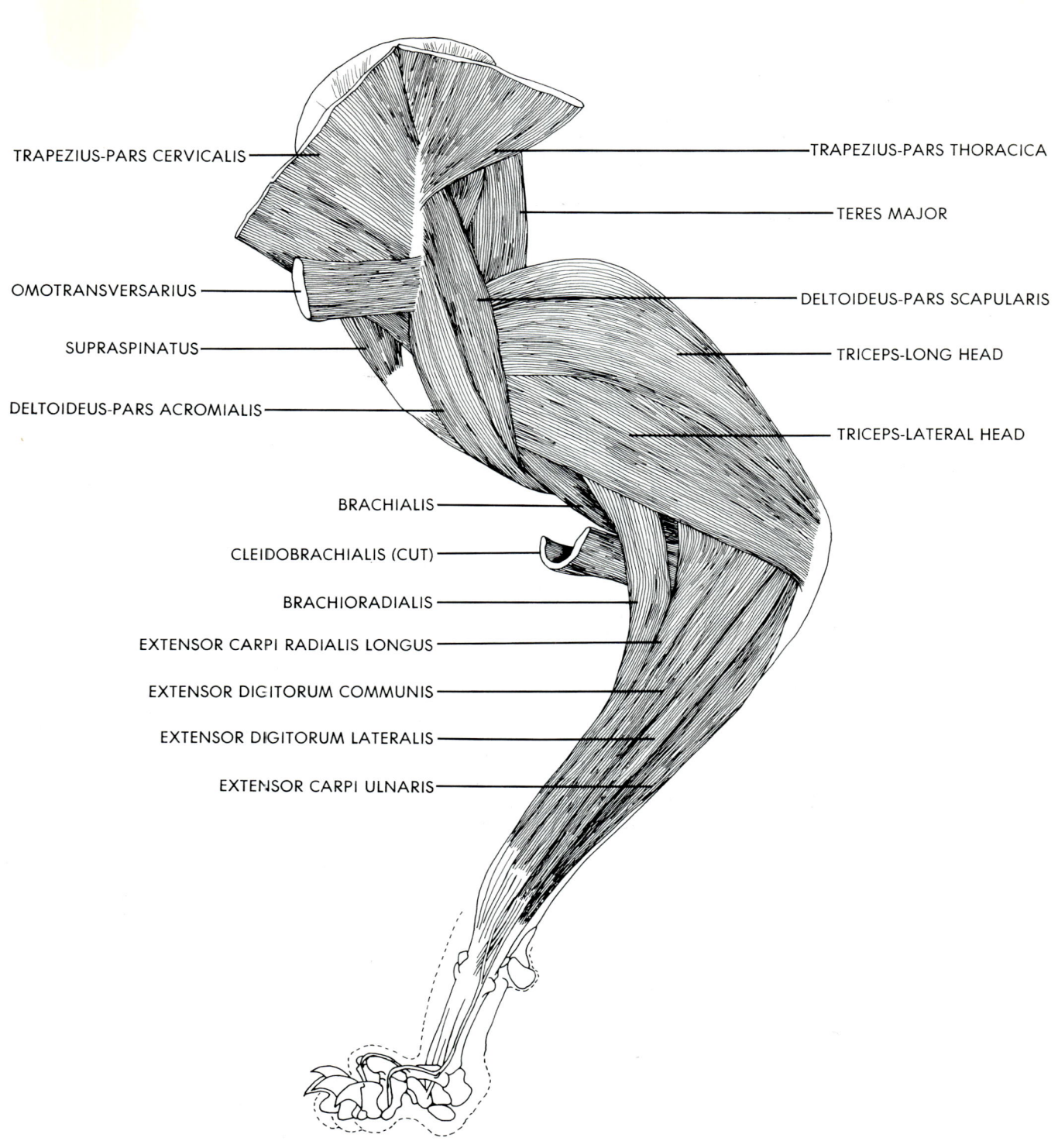

FIGURE 24. Deeper lateral muscles of the shoulder, brachium, and arm. Separate, clean and section the following muscles to expose the muscles illustrated in fig. 25: *Deltoideus-pars acromialis, deltoideus-pars scapularis, triceps brachii lateralis* and *longus, cleidobrachialis, brachioradialis, extensor carpi radialis longus, extensor digitorum lateralis,* and *extensor carpi ulnaris.*

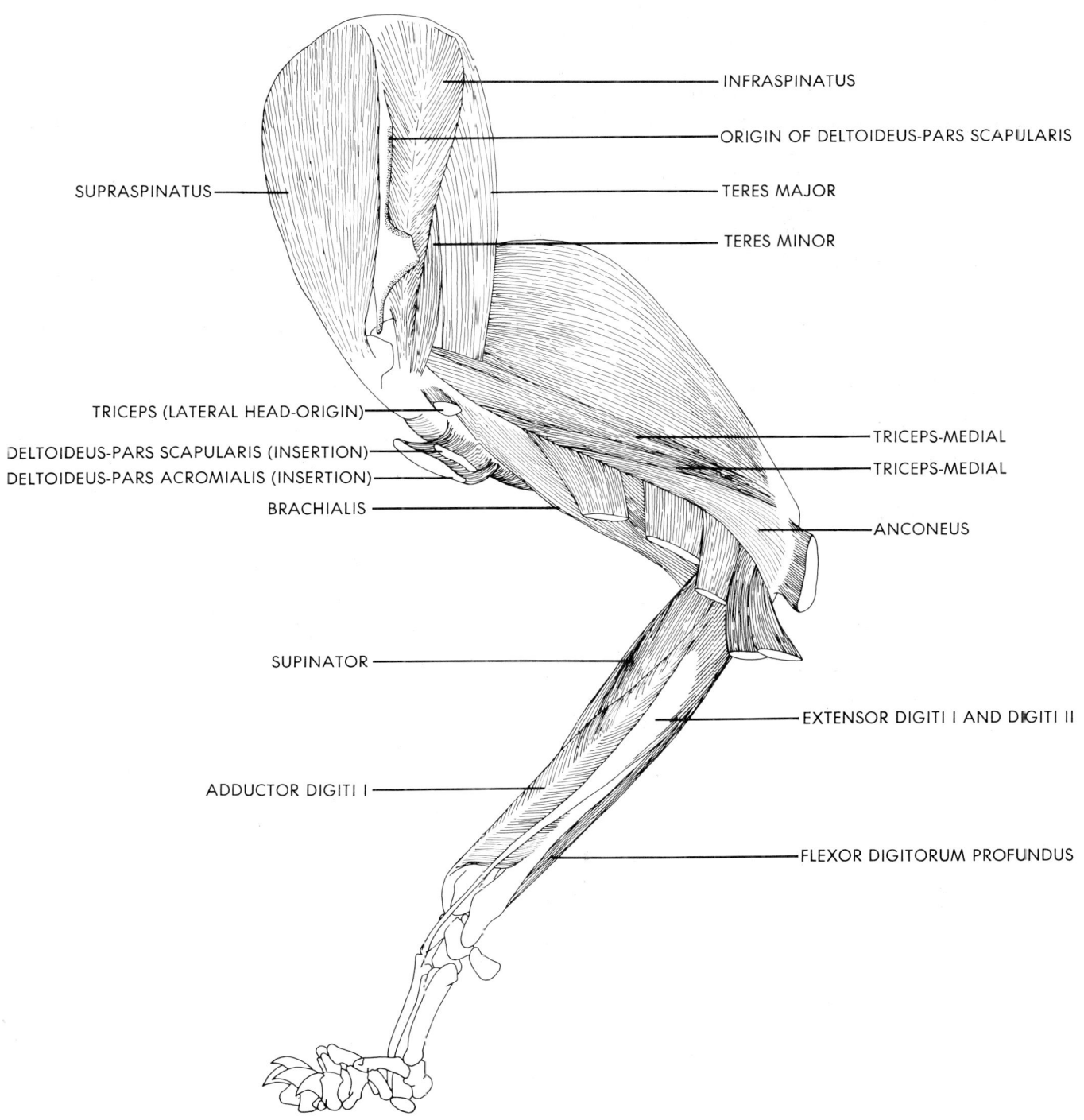

FIGURE 25. The deepest lateral muscles of the shoulder, brachium, and arm. The limb must be removed from the trunk in order to observe the muscles in fig. 27. Observe the muscles attaching the limb to the trunk in fig. 26. Section the following: *Pectoralis superficialis, pectoralis profundus, latissimus dorsi, cleidomastoideus, rhomboideus cervicis, rhomboideus thoracis, serratus ventralis thoracis,* and *serratus ventralis cervicis.*

BREAST

Separate the muscles from one another with your blunt probe. Transect the muscle as near its midpoint as you can in order to leave the origin and insertion intact. Transect the *descendens* portion of the *Pectoralis superficialis* first. Separate the *transversus* portion of the superficial pectoralis from the adjacent muscle and transect it. The *Latissimus dorsi* (see fig. 27) attaches in part to the *Pectoralis profundus*. Carefully separate the *Latissimus dorsi* before transecting the *Pectoralis profundus* (both *cranialis* and *caudalis*). Be careful to not cut the blood vessels and nerves in the axilla (see fig. 51 before cutting).

Note the *Pectoralis profundus* appears to have three parts, the most caudal has been termed the 'xiphihumeralis' and appears to go beneath (deep) to the middle portion. Both the middle and 'xiphihumeralis' portions form the caudalis and insert together on the bicipital groove of the humerus. The *cranialis* portion also inserts in the bicipital groove but somewhat separate to the other *caudalis* portions and its insertion is continuous with a tendon that passes to the coracoid process and fuses with the tendon of insertion of the *Supraspinatus*. These divisions were formerly termed separate muscles but the divisions are no longer recognized by the Nomina Anatomica Veterinaria. Transecting the *Pectoralis* will expose the *Transversus costarum* and *Serratus ventralis*. Now turn the cat over and dissect the shoulder muscles. Rostrally clean the *Brachiocephalicus* (= *Cleidocephalicus*) from adjacent muscles and transect it. Next separate the thoracic and cervical portions of the *Trapezius*. This will expose the cervical and thoracic portions of the *Rhomboideus* which should be cleaned and transected. The limb is now attached to the trunk by the *Sternomastoideus*, *Serratus ventralis* (thoracic and cervical portions), *Omotransversarius*, and the *Latissimus dorsi* plus blood vessels and nerves.

NAME	ORIGIN	INSERTION	ACTION
Pectoralis superficialis descendens	Manubrium of sternum.	Superficial fascia of elbow.	Adducts arm.
Pectoralis superficialis transversus	Raphe of midventral line, manubrium, and sternebrae.	Shaft of humerus.	Adducts arm.
Pectoralis profundus			
pars cranialis*	Sternebrae	Bicipital groove of humerus.	Adducts arm.
pars caudalis*	Xiphisternum	Bicipital groove of humerus.	Adducts arm.

*These two parts are not recognized by the N.A.V. but the muscles are distinct and the omission would be confusing.

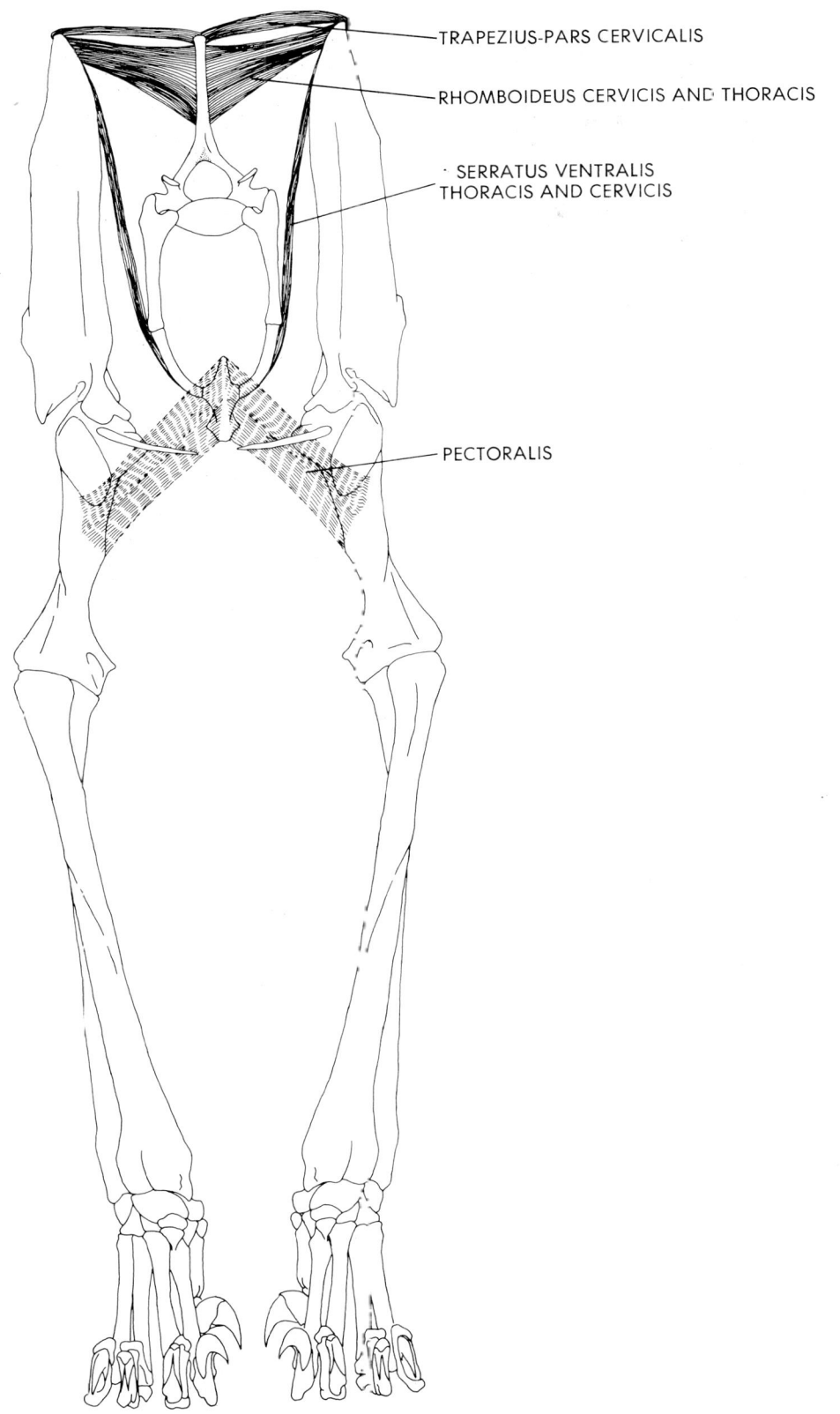

FIGURE 26. Frontal view of the extrinsic shoulder muscles. Compare this with fig. 10. The *serratus ventralis thoracis* and *cervicis* muscles form a 'sling' that serves as a support for the trunk.

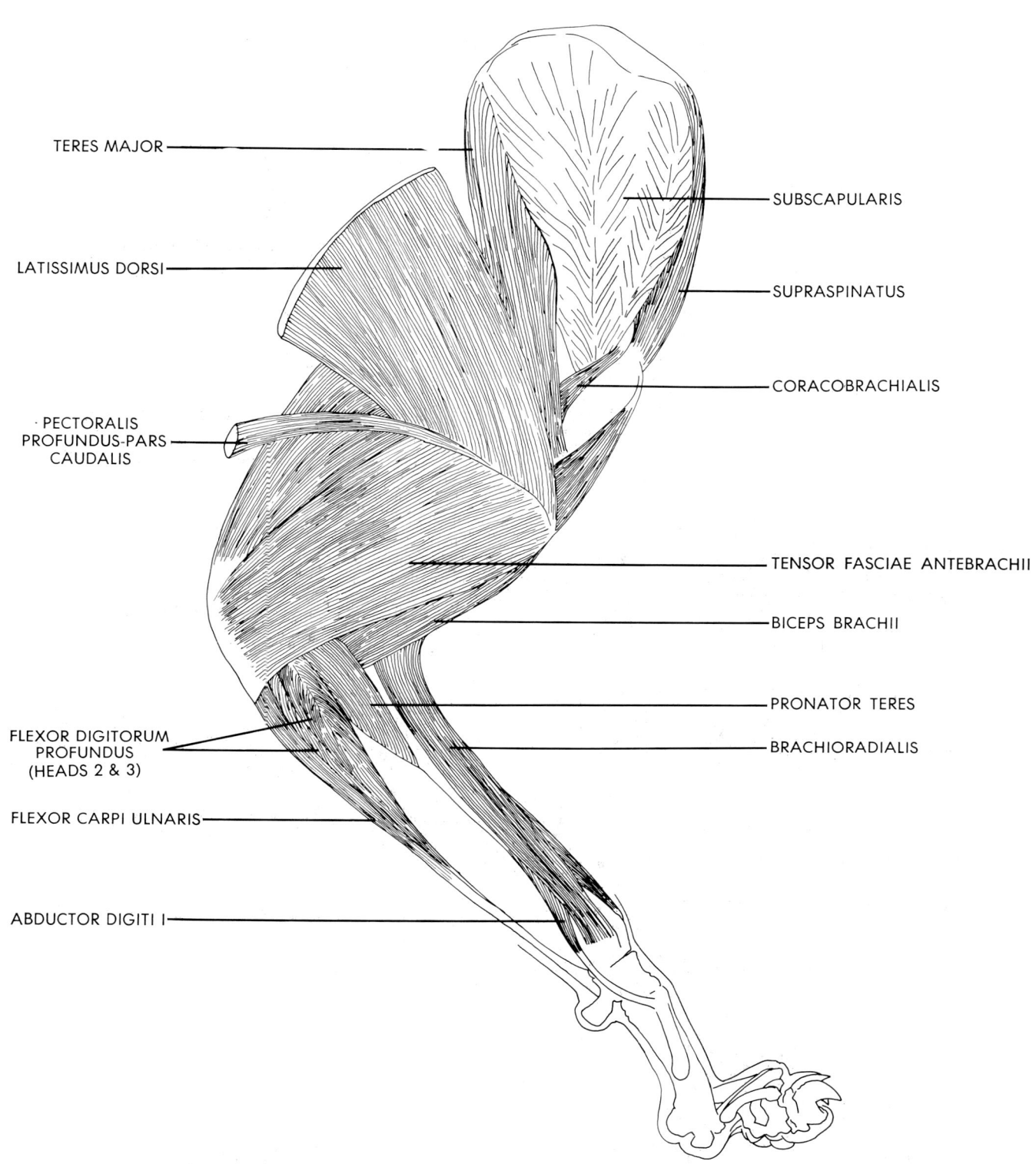

FIGURE 27. Muscles on the medial surface of the pectoral limb. Section *teres major* and remove *latissimus dorsi*, in order to observe the musculature seen in fig. 28. Note: *Brachiocephalicus* and *extensor carpi radialis longus* are illustrated intact in this view although they were actually cut in fig. 25.

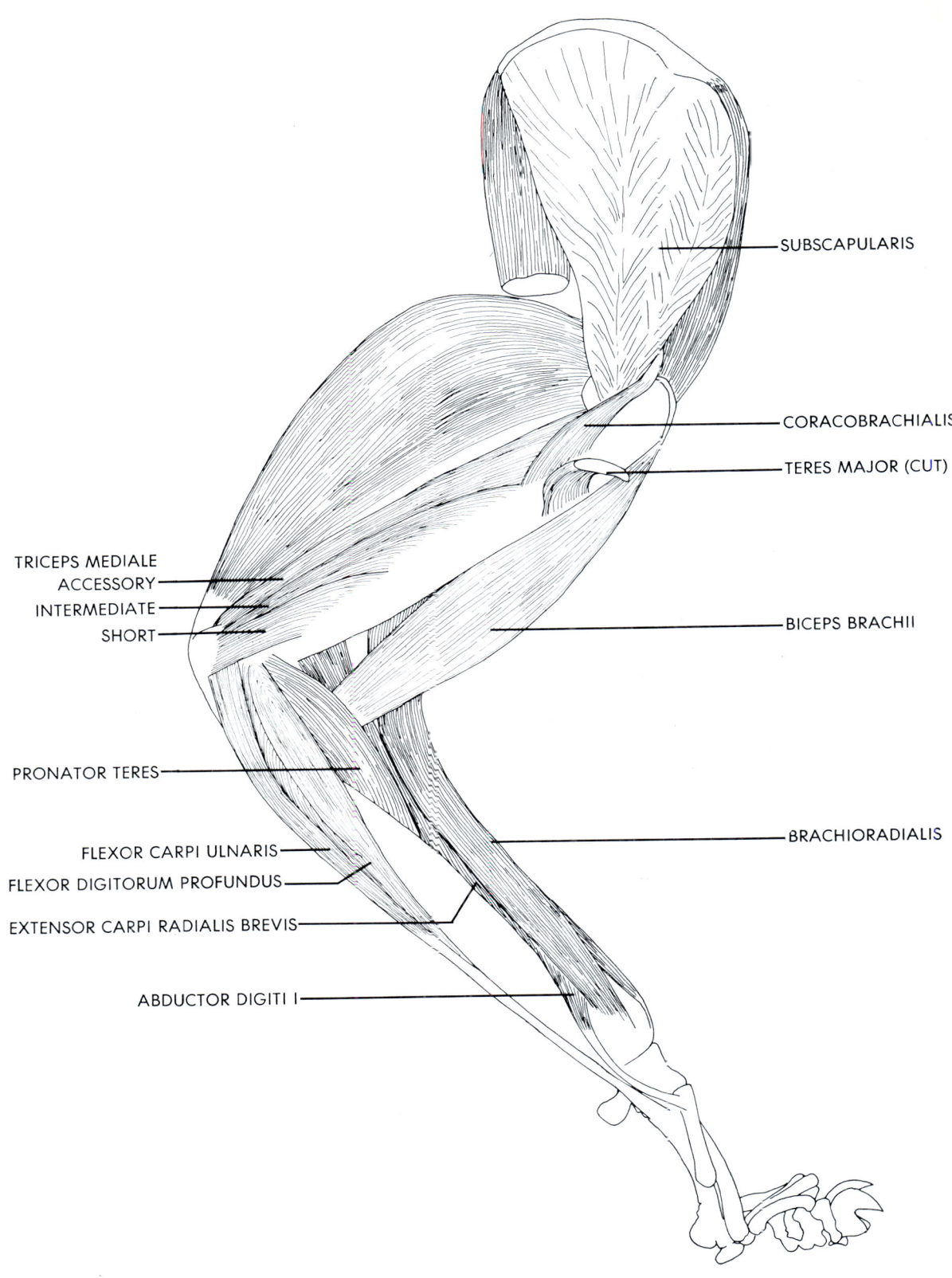

FIGURE 28. Deep medial muscles of the pectoral limb.

ARM (BRACHIUM)

Most of the medial surface of the brachium is covered by the *Tensor fasciae antebrachii*. Loosen and transect this muscle to observe the muscles illustrated in figure 28. Extend the forearm and loosen the *Triceps brachii laterale* with a blunt probe. This will allow you to observe the other parts of the *Triceps brachii* and the *Anconeus*.

NAME	ORIGIN	INSERTION	ACTION
Tensor fasciae antebrachii	Lateral surface of latissimus dorsi.	Olecranon process of humerus.	Extends antibrachium.
Biceps brachii	Bicipital tubercle of scapula.	Bicipital tuberosity of radius.	Flexes antibrachium.
Brachialis	Lateral surface of humerus.	Ulna, distal to semilunar notch.	Flexes antibrachium.
Triceps brachii laterale longum mediale	Deltoid of humerus. Glenoid border of scapula. Shaft of humerus.	Olecranon of ulna.	Extends antibrachium.
Anconeus	Dorsal surface of humerus.	Ulna, lateral to semilunar notch.	Fixes the elbow joint.

FOREARM (ANTIBRACHIUM)

Cut the *Brachioradialis*, *Extensor carpi radialis longus*, *Extensor carpi radialis brevis*, *Extensor digitorum communis*, *Extensor digitorum lateralis* and *Extensor carpi ulnaris* to observe the deeper forearm muscles.

NAME	ORIGIN	INSERTION	ACTION
Brachioradialis	Dorsal border of humerus.	Styloid process of radius.	Supinates hand.
Extensor carpi radialis longus	Supracondyloid ridge of humerus.	Second metacarpal.	Extends hand.
Extensor carpi radialis brevis	Supracondyloid ridge of humerus.	Third metacarpal.	Extends hand.
Extensor digitorum communis	Supracondyloid ridge of humerus.	Second phalanx of all digits except first.	Extends digits.
Extensor digitorum lateralis	Supracondyloid ridge of humerus.	With extensor digitorum communis to second phalanx except 1st and 2nd digits.	Extends digits.

NAME	ORIGIN	INSERTION	ACTION
Extensor carpi ulnaris	Lateral epicondyle of humerus and ulna near semilunar notch.	Base of fifth metacarpal.	Extends wrist.
Extensor digiti II	Ulnar near semilunar notch.	Second phalanx of 2nd digit.	Extend 2nd (index) digit.
Supinator	Ulna and interosseus ligaments.	Proximal and medial surface of radius.	Supinates hand.
Extensor digiti I	Shaft of radius and ulna and interosseus membrane.	Base of 1st metacarpal.	Extends and abducts pollex.
Pronator teres	Medial epicondyle of humerus.	Medial border of radius.	Pronate hand.
Flexor carpi radialis	Medial epicondyle of humerus.	Base of 2nd and 3rd metacarpals.	Flexes wrist.
Palmaris brevis	Medial epicondyle of humerus.	Tendon pad which inserts at base of each digit.	Flexes digits.
Flexor carpi ulnaris caput humerale caput ulnare	Humerus near medial epicondyle. Olecranon process of ulna.	Accessory carpal bone. Accessory carpal bone.	Flexes wrist. Flexes wrist.
Flexor digitorum superficialis	Insertion tendon of palmaris brevis and tendon of flexor digitorum profundus.	Second phalanx of 2nd, 3rd, 4th, and 5th digits.	Flexes digits.
Flexor digitorum profundus	Ulna near semilunar notch. Medial epicondyle of humerus. Radius, ulna, and interosseus membrane.	Last phalanx of each digit.	Flexes digits.
Pronator quadratus	Ulna and interosseus membrane.	Ventral surface of radius.	Pronates hand.

HAND

NAME	ORIGIN	INSERTION	ACTION
Lumbricales	Tendon of flexor digitorum profundus.	Base of 1st phalanx.	Draws digits medially.
Abductor digiti I	Ligament over pisiform bone.	First phalanx of pollex.	Abducts thumb.
Flexor digiti I brevis	Os magnum and scapholunar bones.	First phalanx of pollex.	Abducts thumb.
Adductor digiti I	Os magnum.	First phalanx of pollex.	Adducts thumb.
Interossei	Ventral surface of metacarpals.	First phalanx of digits.	Flexes digits.
Flexor digiti II (Interosseus of second digit)	Ventral surface of 2nd metacarpal.	First phalanx of 2nd digit.	Flexes index finger.
Abductor digiti II	Ventral surface of 2nd metacarpal and trapezium.	First phalanx of 2nd digit.	Abducts index finger.
Adductor digiti II	Os magnum.	First phalanx of 2nd digit.	Adducts index finger.
Abductor digiti V	Pisiform.	First phalanx of 5th digit.	Abducts 5th digit.
Flexor brevis digiti V	Fifth metacarpal and unciform.	First phalanx of 5th digit.	Flexes 5th digit.
Adductor digiti V	Os magnum.	Fifth metacarpal and 1st phalanx of 5th digit.	Adducts 5th digit.

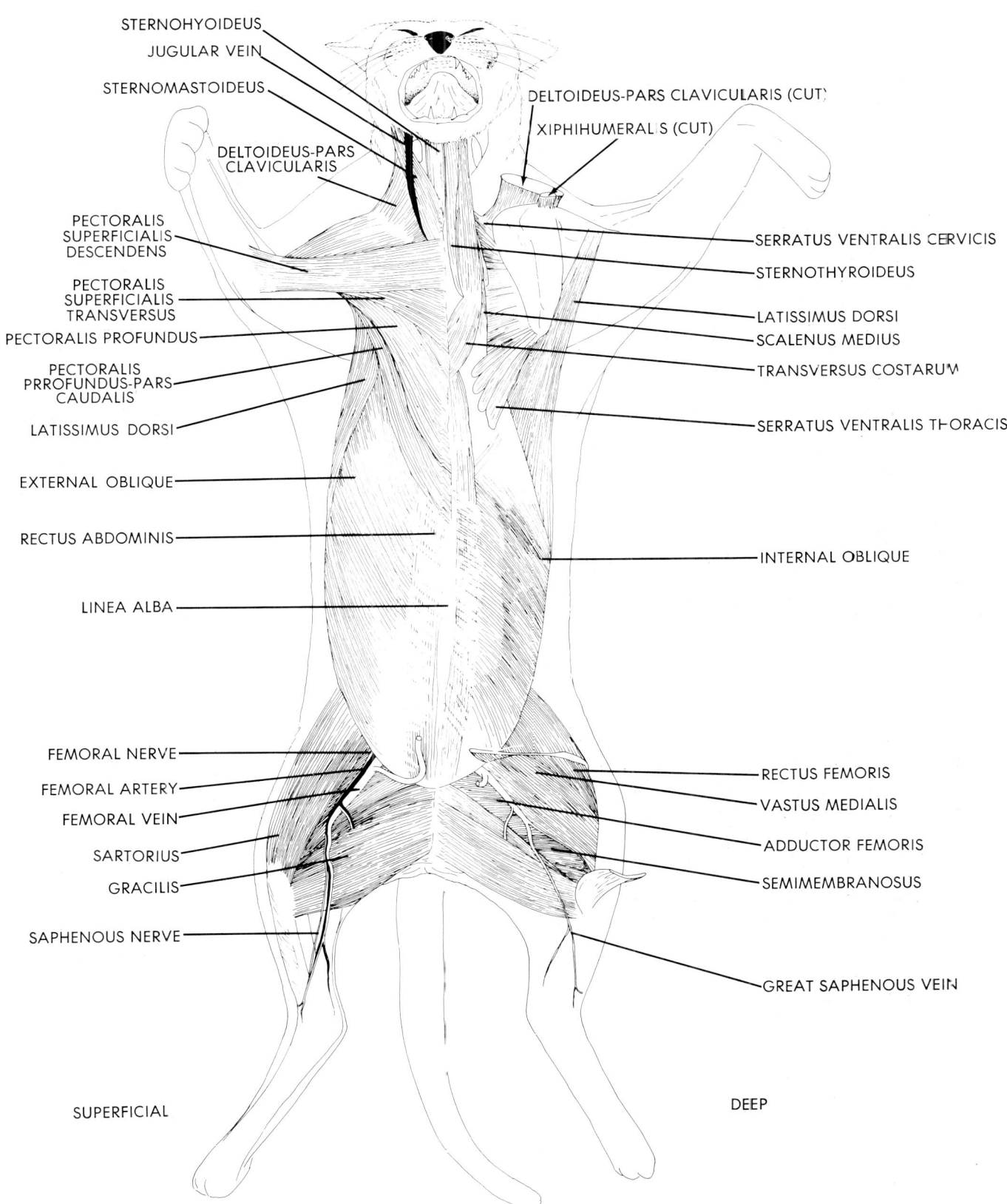

FIGURE 29. Ventral view of the superficial and deep musculature of the cat.

BACK AND TRUNK

The caudal back muscles are covered by the heavy lumbrodorsal fascia. Pinch the lumbrodorsal fascia with forceps and cut through the fascia lateral to the midline without cutting the muscle beneath the fascia. Continue to cut away the fascia laterally until the fascia passes deep to the lateral border of the *Longissimus lumborum*. Grasp the lateral cut border of the fascia with forceps and cut the fascia lateral to this deep fascia to expose another muscle, the *Iliocostalis lumborum*. Note the *Serratus dorsalis caudalis* originates on the fascia covering the *Iliocostalis* and the *Spinalis thoracis* and *cervicis* originates on the rostral portion of the *longissimus fascia*.

The *Serratus dorsalis cranialis* originates from a sheet of fascia over the dorsal thoracic region. Cut the sheet of fascia so the origin of *Serratus dorsalis cranialis* is not disturbed. The *Spinalis thoracis* and *cervicis* and *Longissimus capitis* and *cervicis* are now exposed.

NAME	ORIGIN	INSERTION	ACTION
Longissimus dorsi			
Longissimus lumborum	Iliac crest and articular processes of lumbar vertebrae.	Mammillary and accessory processes of lumbar vertebrae.	Extends lumbar vertebrae.
Longissimus thoracis	Spinous and mammillary processes of lumbar vertebrae.	Accessory processes of thoracic vertebrae and ribs.	Extends thoracic vertebrae.
Longissimus cervicis	Transverse and articular processes of thoracic vertebrae.	Transverse processes of cervical vertebrae.	Extends cervical vertebrae.
Longissimus capitis	Articular processes of last 4 cervical vertebrae by long tendons.	Mastoid process of temporal.	Turns head latterly.
Iliocostalis	Ilium, deep lumbar fascia, transverse processes of lumbar vertebrae and ribs.	Ribs and transverse processes of lumbar vertebrae cranial to origin.	Draws ribs caudally.
Serratus dorsalis			
pars caudalis	Lumbrodorsal fascia covering Longissimus lumborum.	Last 4 ribs.	Draws ribs caudally.
pars cranialis	Fascia covering Spinalis muscles.	First 9 ribs.	Draws ribs cranially.
Spinalis dorsi			
pars thoracis	Lumbrodorsal fascia covering Longissimus lumborum thus to the spinous processes of lumbar and thoracic vertebrae.	Spinous processes of the first 10 thoracic vertebrae.	Extends vertebral column.
pars cervicis	Spinous processes of thoracic vertebrae.	Spinous processes of all cervical vertebrae except the atlas.	Extends neck.
Splenius	First 2 thoracic vertebrae and the raphe in the midline between the two splenius muscles. Beneath fascia of origin of Serratus dorsalis cranialis.	Nuchal crest of the occipital bone.	Raises and turns head laterally.

NAME	ORIGIN	INSERTION	ACTION
Multifidi	Lamina and transverse processes of sacral, lumbar, and thoracic vertebrae.	Spinous processes of vertebrae cranial to origin.	Extends vertebral column.
Interspinales	Spinous process of vertebrae.	Spinous process of vertebrae.	Draw vertebral spines together.
Intertransversarii	Transverse process of vertebrae.	Transverse process of vertebrae.	Draw transverse processes together.
Transversus costarum	Sternum between 3rd and 6th ribs.	First rib and cartilage.	Draws sternum forward (holds sternum in place).
Levatores costarum	Transverse processes of thoracic vertebrae.	Angle of rib.	Draws ribs forward.
Intercostales externi	Ribs cranial to insertion, near vertebrae.	Ribs caudal and ventral to origin.	Draws ribs forward.
Intercostales interni	Ribs and cartilages near sternum ventral and cranial to insertion.	Ribs and cartilages caudal to origin.	Draws ribs forward.
Transversus thoracis	Dorsal sternum.	Cartilage of ribs.	Draws ribs forward.
Diaphragm (figs. 42, 64, 65)	Ventral surface of centrum of 2nd, 3rd, and 4th lumbar vertebrae and xiphoid process of sternum and last 5 ribs.	Central semilunar tendon.	Depresses abdominal viscera and expands thoracic cavity.
Obliquus externus abdominis	Last 9 or 10 ribs, and lumbrodorsal fascia.	Ventral raphe (linea alba) *via* aponeurosis.	Constrict abdomen.
Obliquus internus abdominis	Lumbrodorsal fascia and crural arches between crest of ilium and pubic spine.	Ventral raphe (linea alba) *via* aponeurosis.	Constrict abdomen.
Transversus abdominis	Rib cartilages, transverse processes of lumbar vertebrae, ilium and crural arches between ilium and pubis.	Ventral raphe (linea alba) *via* aponeurosis.	Constrict abdomen.
Rectus abdominis	Tubercle of pubis.	First and 2nd rib cartilages.	Compresses abdomen, and maintains vertebral column arch.

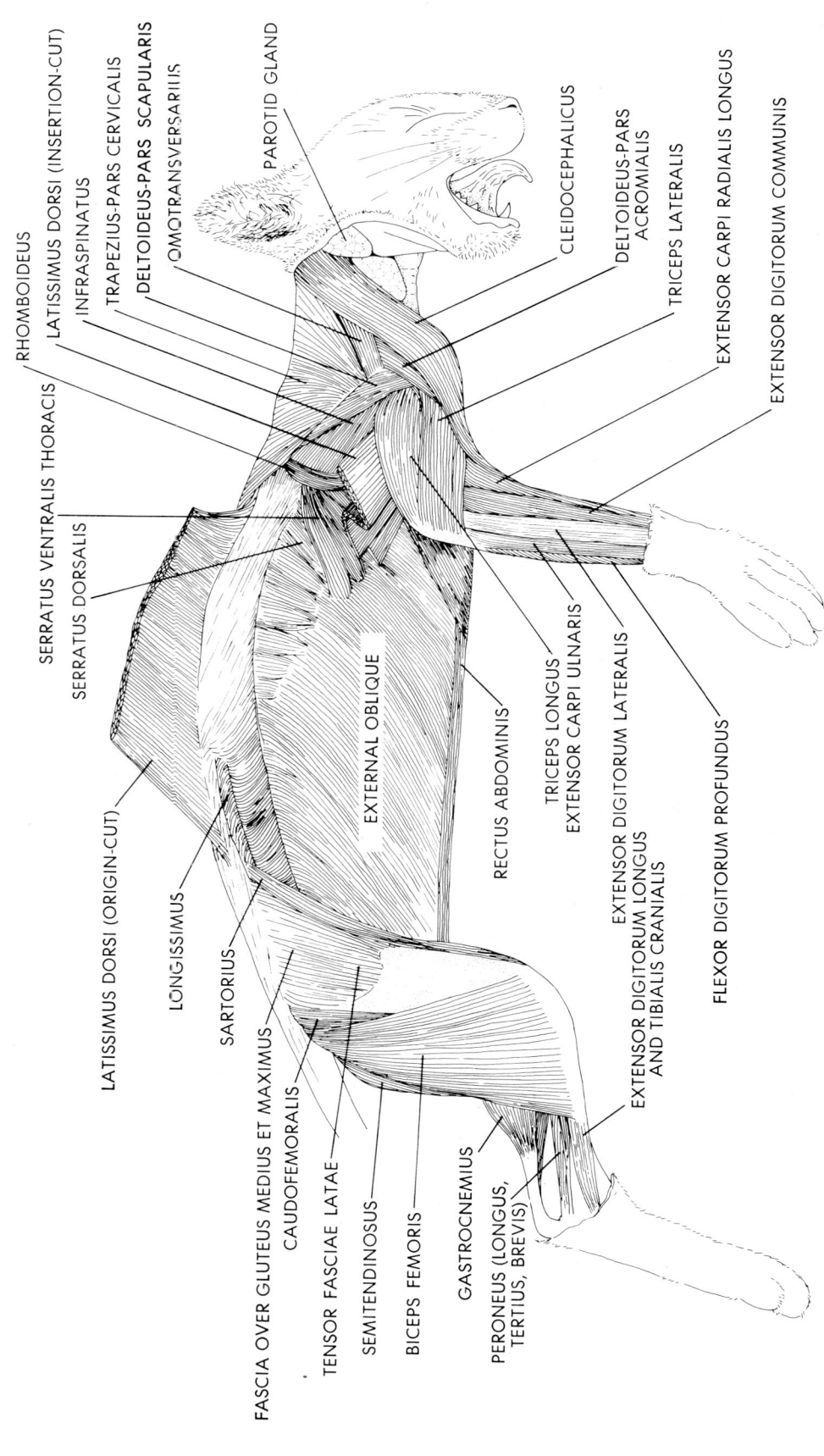

FIGURE 30. Lateral view of the superficial trunk and limb musculature.

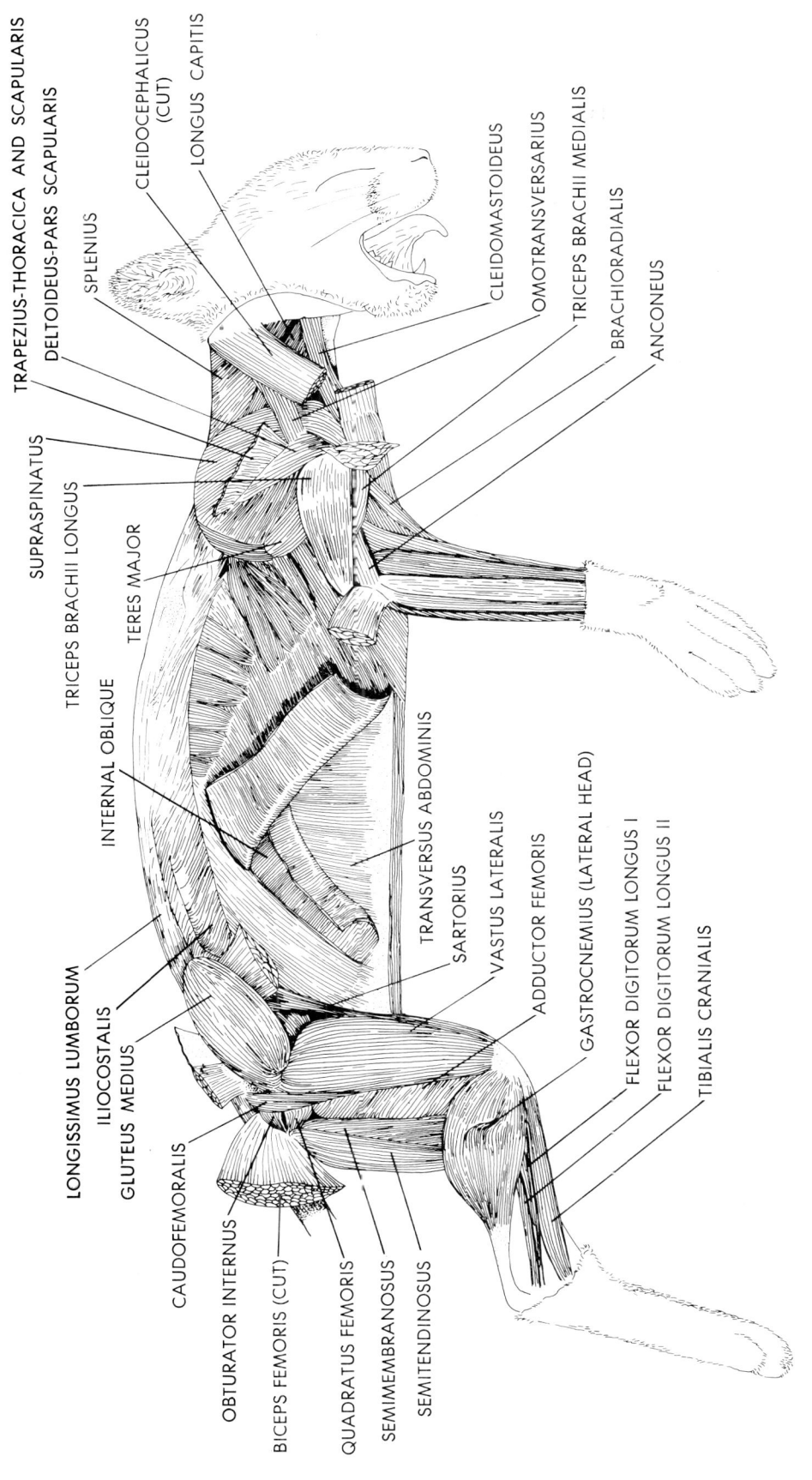

FIGURE 31. Lateral view of the deep trunk and limb musculature.

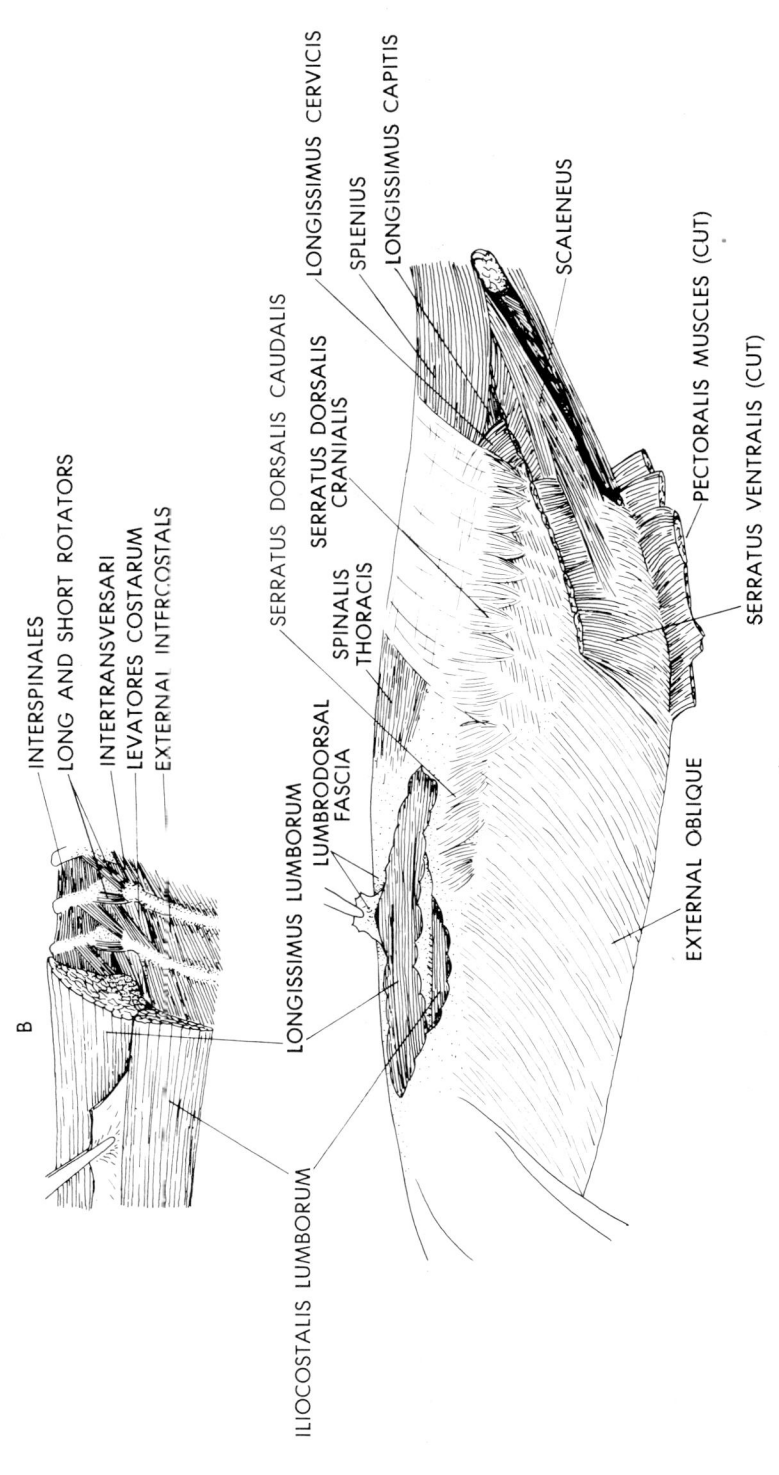

FIGURE 32. Lateral views of the deep trunk and vertebral muscles. A. Superficial trunk muscles with the lumbrodorsal fascia cut and reflected. B. Detail of the deeper vertebral muscles.

DEEP NECK MUSCLES
Ventral

The midline neck muscles beneath the trachea and esophagus are the longus colli muscles from the thorax to the axis. The rectus capitis ventralis fibers run in the same oblique direction as the longus colli fibers and just rostral to the longus colli, inserting on the atlas. The longus capitis is lateral to the longus colli.

NAME	ORIGIN	INSERTION	ACTION
Longus colli	Transverse process of first 6 thoracic and all cervical vertebrae.	Ventral centrum of all cervical vertebrae.	Arches neck (with head down).
Rectus capitis ventralis	Ventral transverse process of atlas.	Lateral to occipital condyle.	Flexes head laterally.
Longus capitis	Ventral transverse processes of 2nd to 6th cervical vertebrae.	Basioccipital.	Depresses snout.

Lateral

The scaleneus muscles originate on either side of the insertion of the serratus ventralis muscles (page 50 and fig. 34) and insert on the transverse processes of the last four cervical vertebrae. The longissimus cervicis insertion (see BACK AND TRUNK, page 42) is just dorsal to the insertion of the scaleneus. The longissimus capitis (BACK AND TRUNK) extends forward to the head (mastoid process) just dorsal to the insertion of the longissimus cervicis. Cut the longissimus capitis and reflect toward its origin and insertion.

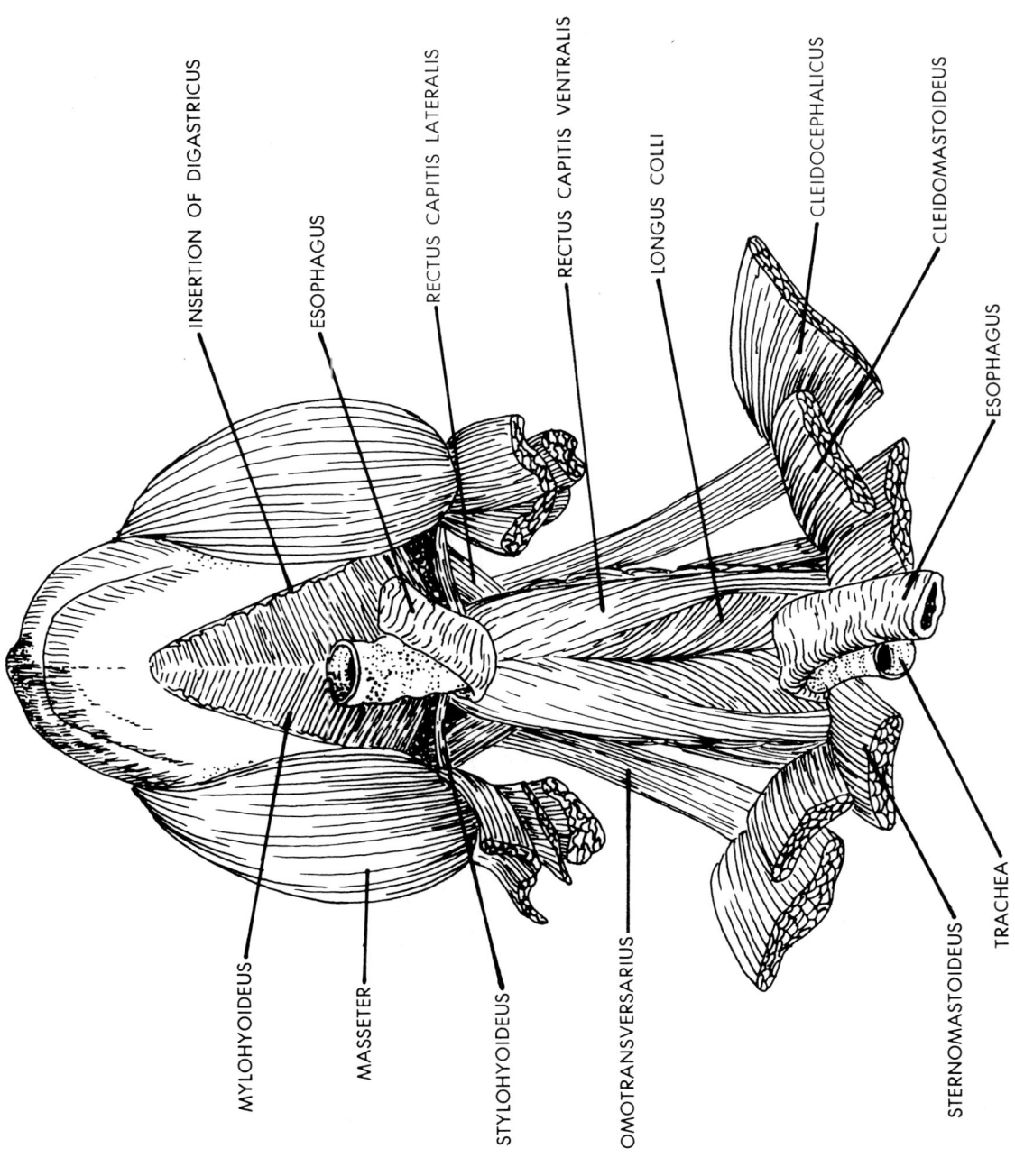

FIGURE 33. Ventral view of the deep neck muscles. Compare with fig. 22.

DEEP NECK MUSCLES (Continued)

The splenius (page 42) covers the dorsal half of the neck. Cut and reflect the splenius. The semispinalis capitis (in two parts, biventer cervicis and complexus) lies deep to the splenius. After examining the origins and insertions cut and reflect both parts of the semispinalis capitis.

NAME	ORIGIN	INSERTION	ACTION
Scalenus			
dorsalis	Second and 3rd ribs.	Transverse processes of 4th to 7th cervical vertebrae.	Flexes neck. May draw ribs forward.
medius	Sixth, 7th, 8th and 9th ribs.		
ventralis	Third or 4th rib.		
Semispinalis			
semispinalis cervicis	Transverse and articular processes of cervical vertebrae.	Spinous processes of cervical vertebrae.	Extends vertebral column.
semispinalis capitis biventer cervicis complexus	Transverse processes of cranial, thoracic, and caudal cervical vertebrae.	Occipital bone.	Extends vertebral column.
Rectus capitis dorsalis major	Spinous process of axis.	Lambdoidal ridge.	Raises snout.
Rectus capitis lateralis	Spine of axis.	Occipital.	Raises snout.
Rectus capitis dorsalis minor	Ventral surface of atlas.	Basioccipital.	Depresses snout.
Obliquus capitis cranialis	Crest of axis.	Transverse process of atlas.	Rotates head.
Obliquus capitis caudalis	Transverse process of atlas.	Mastoid process and occipital below lambdoidal ridge.	Flex head laterally.

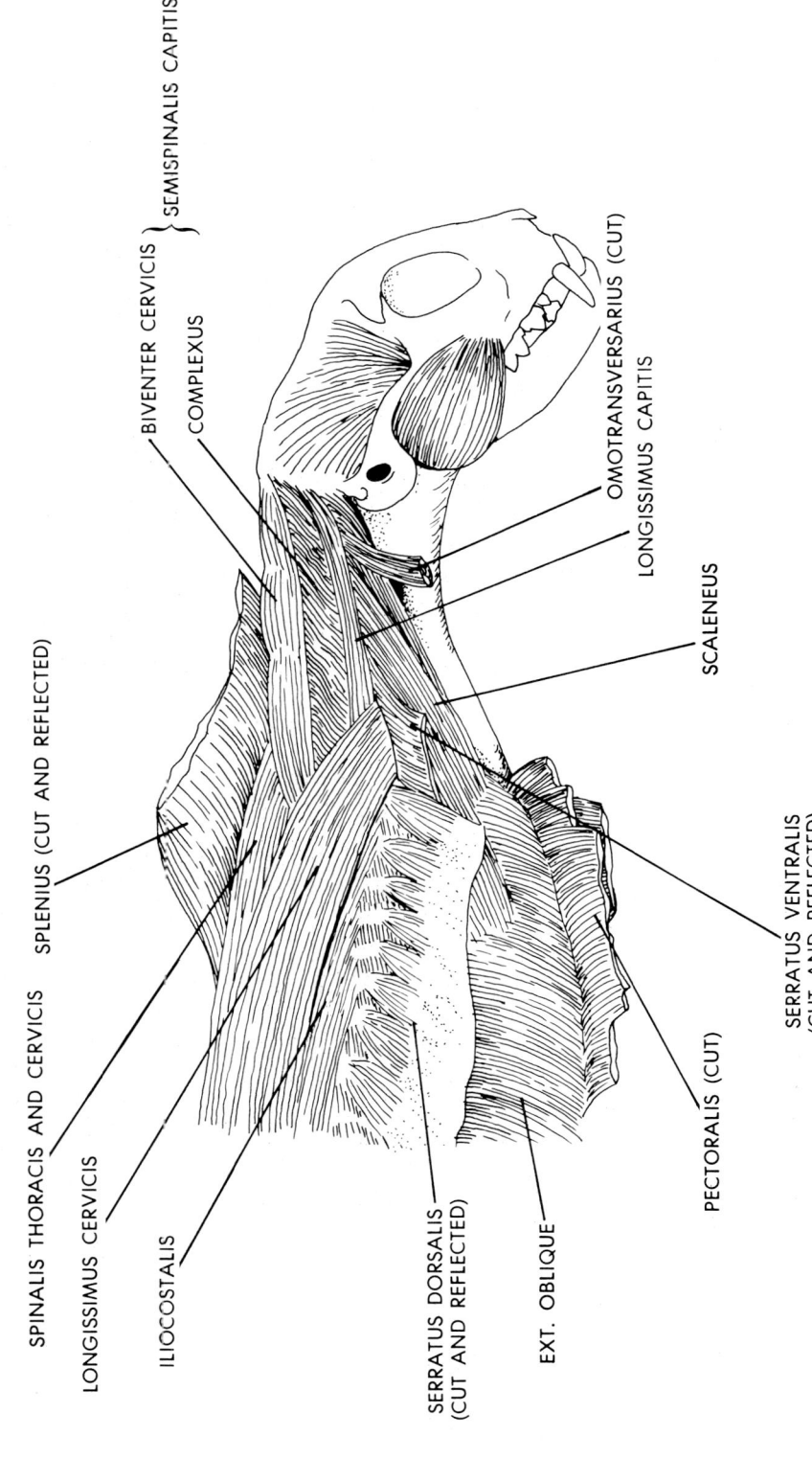

FIGURE 34. Lateral view of the deep neck muscles. Compare with fig. 32.

PELVIC GIRDLE, HIP AND THIGH

Clean, separate and section the *Sartorius* and *Biceps femoris*. Be extremely careful to separate and avoid cutting the slender *Abductor cruris caudalis* which lies just beneath the *Biceps femoris*. Separate the *fascia lata* from the underlying *Vastus lateralis* and *Caudofemoralis* muscles and cut the tendon (*fascia lata*) rather than the muscle (*Tensor fascia lata*).

Examine the muscles in figure 36 and then separate and transect the *Semitendinosus, Semimembranosus, Quadratus femoris, Caudofemoralis, Gluteus medius* (remove the origin of this muscle instead of transecting it), *Gluteus superficialis*, and the *Abductor cruris caudalis*. Gently extend the shank and put the tip of your blunt probe between the *Vastus lateralis* and the femur near the muscle origin. Now draw your probe toward the hip thus separating the muscle from the femur at its origin. With the cat lying on its back gently probe the small triangular gap at the middle inguinal portion of the thigh. This probing should expose the femoral artery, vein, and nerve. Pick the fat away from this area without injuring the vessels or nerve and then separate the muscles with a blunt probe. The *Pectineus* lies just caudal to the vessels and nerve and the *Adductor longus* is just caudal to the *Pectineus*. The loin muscles, *Psoas major, Psoas minor, Iliacus*, and *Quadratus lumborum* insert on the femur just cranial to the blood vessels and nerve. Note that these muscles emerge from the abdominal cavity through an arch of tendonous sheaths in the iliopectineal region. The most cranial arch is a passageway for the femoral nerve and the loin muscles and the caudal arch allows passage of the femoral artery and vein. Insert a blunt tip scissors in the cranial arch and cut the abdominal wall forward (toward the sternum) to expose the bellies of the loin muscles.

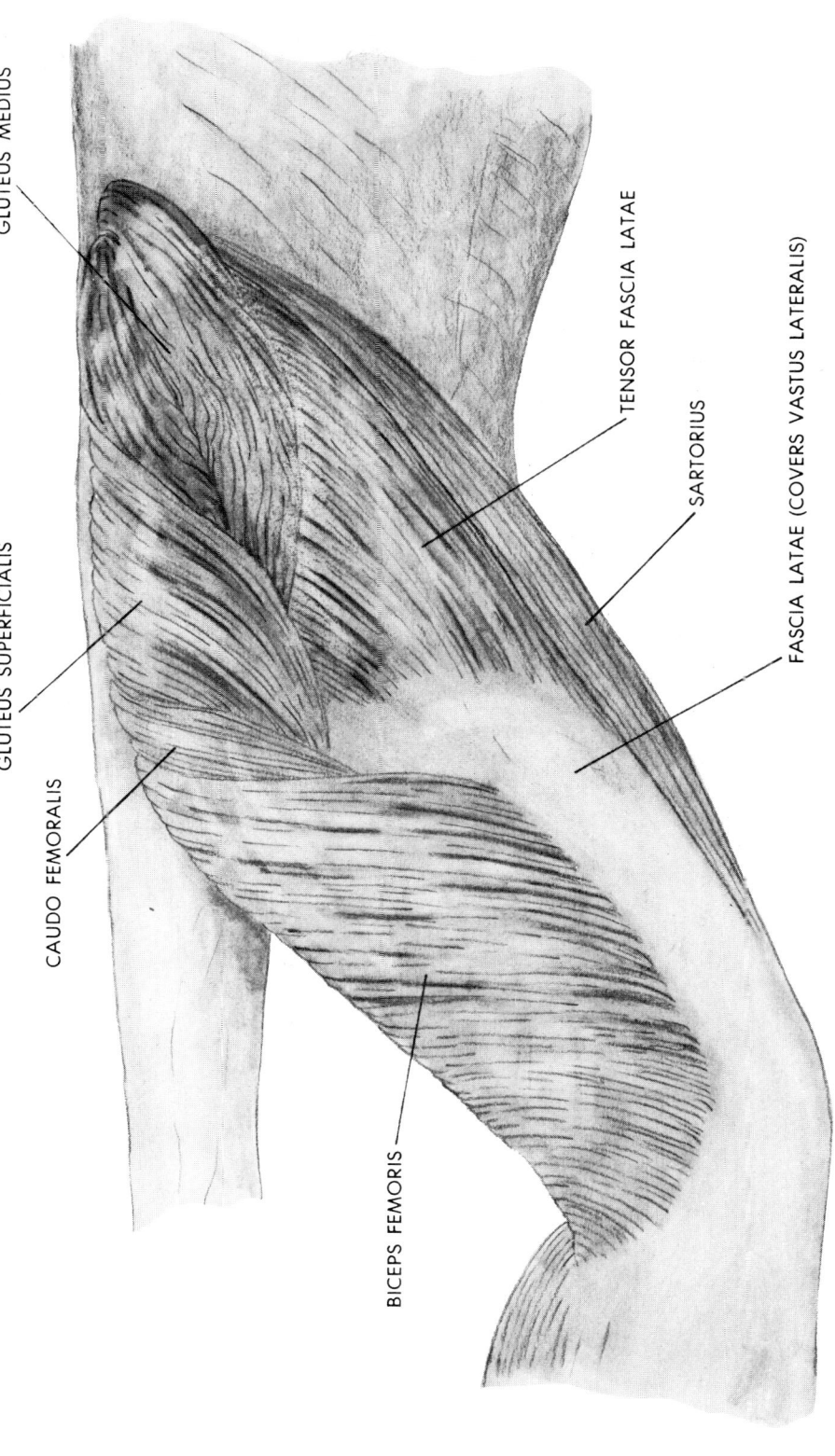

FIGURE 35. Superficial muscles of the hip and thigh. Lateral view.

PELVIC GIRDLE AND HIP

NAME	ORIGIN	INSERTION	ACTION
Psoas minor	Centra of last thoracic and first 4 lumbar vertebrae.	Iliopectineal line cranial to acetabulum of ilium.	Flex lumbar vertebrae.
Quadratus lumborum	Ventral surface of last 2 thoracic vertebrae and last rib.	Transverse processes of lumbar vertebrae and ilium.	Bends vertebral column laterally.
Tensor fasciae latae	Ilium and fascia of gluteus medius.	Fascia lata, connective tissue covering the gluteus medius and vastus lateralis.	Tenses fascia latae and extends leg.
Gluteus superficialis	Transverse processes of last sacral and 1st caudal vertebrae and fascia of gluteus medius.	Great trochanter.	Abducts and extends thigh.
Gluteus medius	Sacral and caudal vertebrae, ilium and fascia of hip region.	Great trochanter.	Abducts and extends thigh.
Gluteus profundus	Ventral half of ilium.	Great trochanter.	Rotates thigh.
Piriformis	Transverse processes of last 2 sacral and 1st caudal vertebrae.	Great trochanter.	Abducts and extends thigh.
Gemellus cranialis*	Dorsal border of ilium and ischium.	Great trochanter.	Rotates and abducts femur.
Gemellus caudalis*	Dorsal border of ischium caudal to Gemellus cranialis.	Great trochanter.	Rotates and abducts femur.
Articularis coxae	Ilium.	Femur, below great trochanter.	Rotates thigh.
Tendon of flexor digitorum	Ischium.	Tendon of obturator internus. Medial surface.	Abducts thigh.
Quadratus femoris	Ischium near tuberosity.	Greater and lesser trochanters.	Extends thigh.
Obturator externus	Pubis and ischium and fascia covering medial surface of obturator foramen.	Trochanteric fossa of femur.	Rotates and flexes thigh.
Obturator internus	Ischium.	Trochanteric fossa of femur.	Abducts thigh.
Iliopsoas (= Iliacus + Psoas)	Psoas minor, lumbar vertebrae, and ventral border of ilium. (Be careful not to destroy the nerve plexus beneath this muscle.)	Lesser trochanter of femur.	Rotates and flexes thigh.

*All gemelli fibers are collectively listed as *Mm. gemelli* by N.A.V.

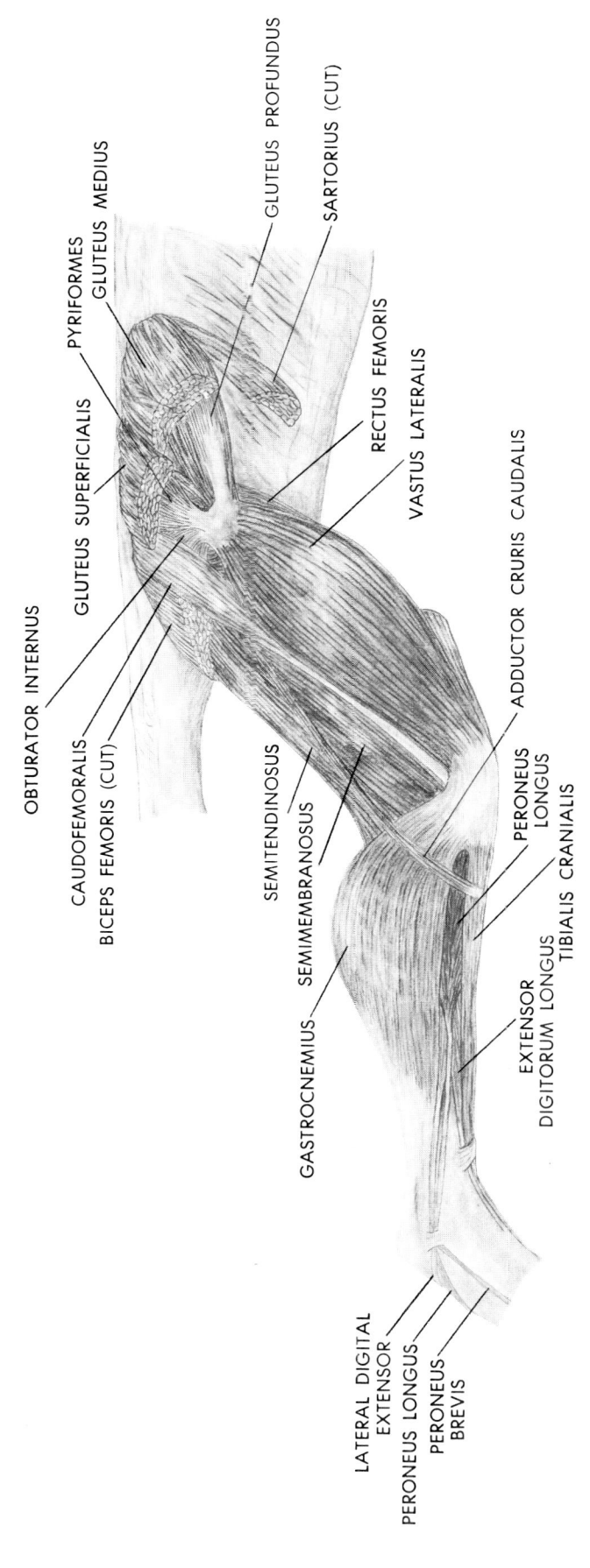

FIGURE 36. Middle layer of muscles of the hip, thigh and shank. Lateral view.

THIGH

NAME	ORIGIN	INSERTION	ACTION
Biceps femoris	Tuberosity of ischium.	Dorsal border of tibia and patella.	Extends and abducts thigh and flexes shank.
Abductor cruris caudalis	Transverse process of second caudal vertebra.	Tendon of insertion of biceps femoris.	Assists biceps.
Caudofemoralis	Transverse process of 2nd and 3rd caudal vertebrae.	Patella.	Abducts and extends thigh.
Semitendinosus	Tuberosity of ischium.	Crest of the tibia.	Extends thigh and flexes shank.
Semimembranosus	Tuberosity and ramus of ischium.	Medial epicondyle of femur.	Extends thigh.
Sartorius	Crest of ilium.	Shaft of tibia.	Adducts femur and extends tibia.
Gracilis	Pubic symphysis and ischium.	Aponeurosis over sartorius and shaft of tibia.	Adducts and flexes shank.
Adductor brevis	Pubis and ischium.	Ventral shaft of femur distal to adductor longus.	Extends thigh.
Adductor magnus	Pubis and ischium.	Shaft of femur distal to adductor brevis.	Extends and adducts thigh.
Adductor longus	Anterior pubis.	Linea aspera of femur.	Extends and adducts thigh.
Pectineus	Anterior pubis.	Shaft of femur near lesser trochanter.	Adducts thigh.
Quadriceps femoris in four parts:			
Rectus femoris	Ventral border of ilium.	Patella.	Extends shank.
Vastus lateralis	Shaft and great trochanter of femur.	Patella.	Extends shank.
Vastus medialis	Shaft of femur.	Patella and head of tibia.	Extends shank.
Vastus intermedius	Dorsal shaft of femur.	Capsule of knee joint.	Extends shank.
Articularis genus	Distal lateral border of femur.	Had of tibia.	Tenses knee ligaments.

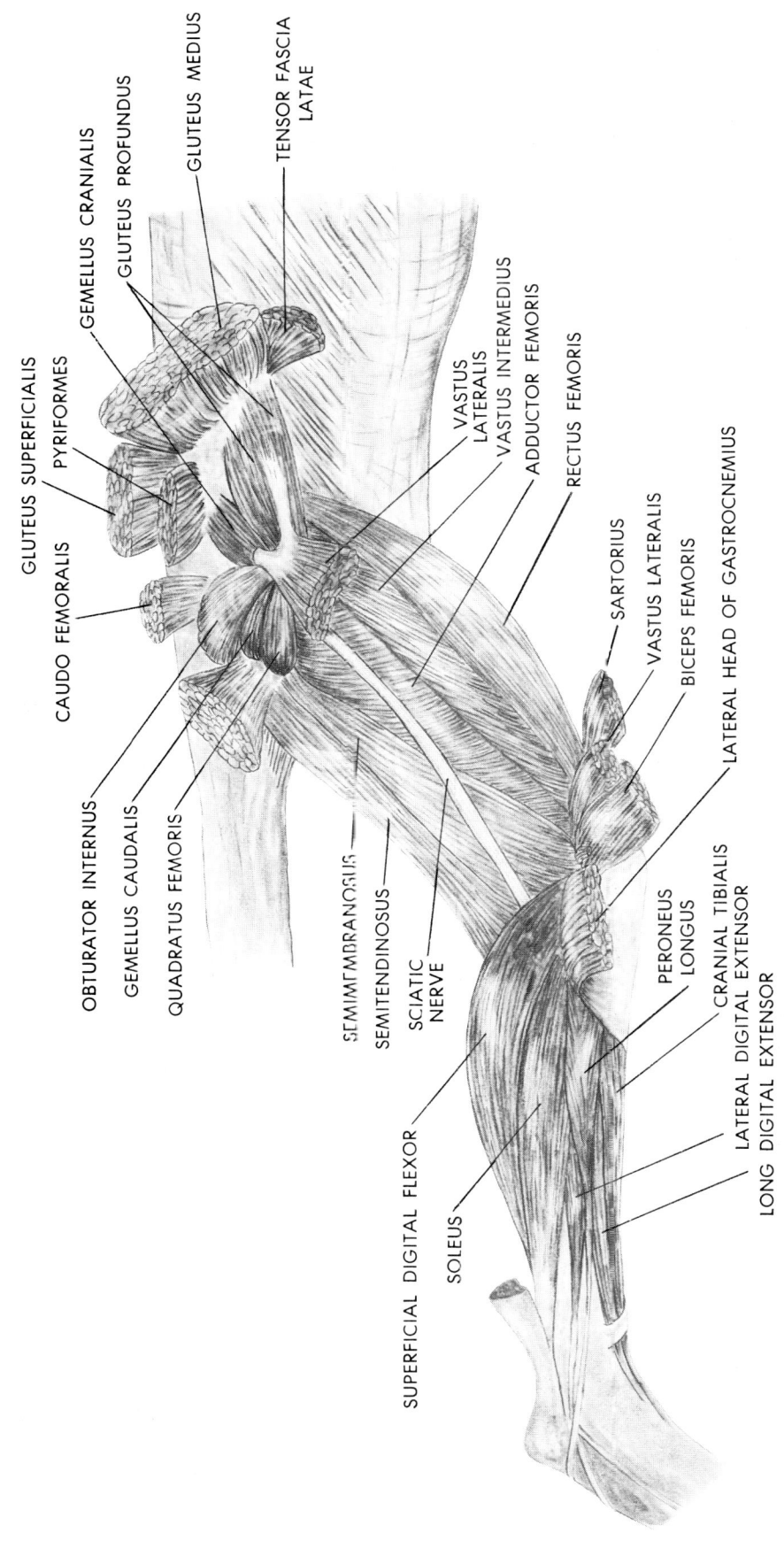

FIGURE 37. Deep muscles of the hip, thigh and shank. Lateral view.

SHANK (Figures 36 and 37)

NAME	ORIGIN	INSERTION	ACTION
Triceps surae			
Gastrocnemius			
laterale	Lateral condyle and superficial fascia of femur.	Tendon of Achilles to calcaneus.	Extends foot.
mediale	Medial epicondyle of femur and shank fascia.	Tendon of Achilles to calcaneus.	Extends foot.
Soleus	Head of fibula.	Tendon of Achilles to calcaneus.	Extends foot.
Superficial digital flexor (= Plantaris)	Patella and lateral condyle of femur with lateral head of Gastrocnemius.	Tendon of Achilles to calcaneus.	Extends foot.
Popliteus	Lateral meniscus of femorotibial joint.	Ventral shaft of tibia.	Rotates thigh and/or shank.
Flexor digiti I longus (hallucis)	Shaft of tibia.	Terminal phalanx of 1st digit.	Flexes big toe.
Flexor digitorum longus	Shaft of tibia and fascia of shank.	Terminal phalanges of each digit.	Flexes toes.

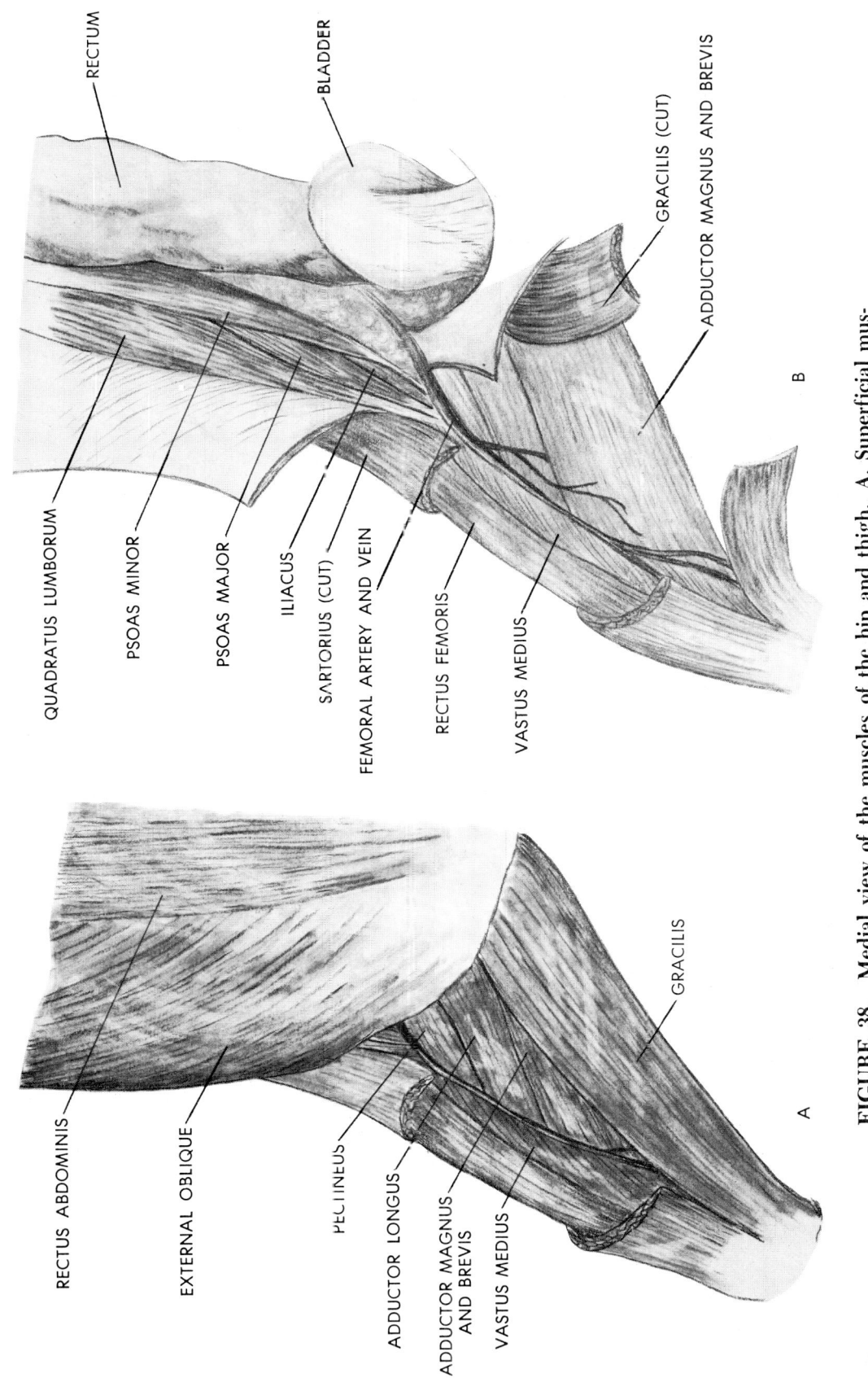

FIGURE 38. Medial view of the muscles of the hip and thigh. A. Superficial muscles with a portion of the *sartorius* removed. B. Deeper muscles with the body cavity opened to show the loin muscles.

NAME	ORIGIN	INSERTION	ACTION
Tibialis caudalis	Medial surface of tibia and fibula near heads.	Lateral tuberosity of scaphoid and medial cuneiform.	Extends foot.
Peroneus longus	Head of fibula.	Base of metatarsals.	Flexes foot.
Peroneus tertius	Lateral surface of fibula.	Extensor tendon of 5th digit.	Extends 5th digit and flexes foot.
Peroneus brevis	Fibula.	Base of 5th metatarsal.	Extends foot.
Extensor digitorum longus	Lateral epicondyle of femur.	By separate tendons to 2nd and terminal phalanges of each digit.	Extends phalanges.
Tibialis cranialis	Lateral shaft of tibia and head of fibula.	First metatarsal.	Flexes foot.

FOOT

NAME	ORIGIN	INSERTION	ACTION
Extensor digitorum brevis	Calcaneus and 3 lateral metatarsals.	First phalanx of each digit.	Extends digits.
Flexor digitorum brevis	Tendon of plantaris insertion.	First and 2nd phalanges and sesamoids of digits.	Flexes 2nd phalanx of digits.
Quadratus plantae	Calcaneus and cuboid.	Tendon of flexor digitorum longus.	Flexes digits.
Lumbricales	Flexor digitorum longus tendon of insertion.	First phalanx of 3rd, 4th, and 5th digits.	Move digits medially.
Interossei	Lateral cuneiform.	First phalanx and tendon of insertion of extensor digitorum longus and extensor digitorum brevis.	Abductors and adductors of digits.
Abductor digiti V	Calcaneus and 5th metatarsal.	First phalanx of 5th digit.	Abducts 5th digit.
Adductor digiti V	Calcaneus and 5th metatarsal.	First phalanx of 5th digit.	Adducts 5th digit.

The muscles of the tail have been omitted.

SUGGESTED READING

Goslow, G. E., Reinking, R. M., and Stuart, D. G. 1973. The cat step cycle: hind limb joint angles and muscle lengths during unrestrained locomotion. *Jour. Morph.* 141(1):1-42.

Kerr, N. S. 1955. The homologies and nomenclature of the thigh muscles of the opossum, cat, rabbit and rhesus monkey. *Anat. Rec.* 121:481-493.

Straus, W. L. Jr., and Sprague, J. M. 1944. Interosseus muscles of the cat. *Amer. Jour. Physiology,* 142.

Chapter 4
Coelomic Membranes and Viscera

This chapter will help orient you to those internal organs described in more detail in subsequent chapters. Although enough information to locate the major organs is given here, the membranes and coelomic cavities are described more completely.

Open the abdominal and thoracic cavities with a scissors cut through the body wall just to the right of the midventral line from the clavicle to the anus. In the thoracic region you should be cutting through the cartilage ribs on the right side. Do not attempt to split the sternum. Insert the blunt tip of your scissors through the muscular body wall in the thorax and keep the sharp tip outside the body. As always, be careful and avoid cutting any internal membranes or organs. Break or cut each rib on the right side near its vertebral articulation. Make lateral incisions where needed and deflect the body wall to the right in order to expose the underlying structures.

Do not make any other cuts or remove any structures until specifically told to do so. Locate the various membranes and structures as best you can without further cutting. The pericardial membranes will be difficult to see at this time; you may observe them better during your study of the circulatory system.

MESENTERIES AND MEMBRANES

Among vertebrates, the coelomic cavities consist of two long tubes on either side of the body midline, analogous to two water-filled balloons with the various visceral organs between them. The membranes of the coelomic cavities are serous membranes: those in the midline, contacting the surface of the visceral organs, are called *visceral;* those contacting the inner surface of the body wall are *parietal.* The double-walled membrane in the midline formed by the contact of membranes from both the right and left coelomic sacs is *mesentery.* Mesentery dorsal to the viscera is *dorsal mesentery;* that ventral to the viscera is *ventral mesentery.*

The first variation of this pattern is seen in fish, in which the cavities and membranes around the heart are "pinched off" from the caudal cavities and both dorsal and ventral mesenteries disappear. The visceral and parietal membranes are all that remain around the coelomic cavity of the fish heart. Since these membranes surround the heart (cardia), they are termed *visceral pericardia* and *parietal pericardia.* The heart and pericardial membranes are located between the two—more caudal—coelomic sacs in the terrestrial vertebrates (reptiles, birds, and mammals). In this arrangement the pericardial sac becomes surrounded by an additional set of coelomic membranes with dorsal and ventral mesenteries. The "new" visceral membrane now surrounds the "old" parietal pericardia. The resulting double membrane surrounding the pericardial coelom is the *pericardium.*

In addition to the heart, the major visceral organs in this region of the mammalian body are the lungs, trachea, and bronchi (see chapter 6), and the esophagus, which passes dorsal to the heart. The heart, esophagus, trachea, and thymus (p. 84), as well as fat and large blood vessels prevent the serosal layers from contacting in the midline to form dorsal and ventral mesenteries. Since they fail to meet, the area between the two medial serosal membranes is termed the *mediastinum.*

The membranes at the caudal end of the pericardial sac become invested with connective tissue, thicken, and form a transverse septum separating the ventral portions of the coelomic cavities into cranial and caudal divisions. In mammals, dorsal muscular folds descend toward and join with the transverse septum to form the *diaphragm.* Although the diaphragm is pierced by the esophagus and nerves and blood vessels, it efficiently separates the coelomic cavities into cranial *pleural* cavities around the lungs and a caudal *peritoneal* cavity surrounding the abdominal viscera.

Thus the coelomic membranes cranial to the diaphragm are either pleural or pericardial and those caudal to the diaphragm are peritoneal.

In the peritoneal region, remnants of the ventral mesentery persist as a *falciform ligament* from the ventral body wall to the liver; the *lesser omentum* from the liver to the lesser curvature of the stomach; and, at the caudal end of the peritoneal cavity, the

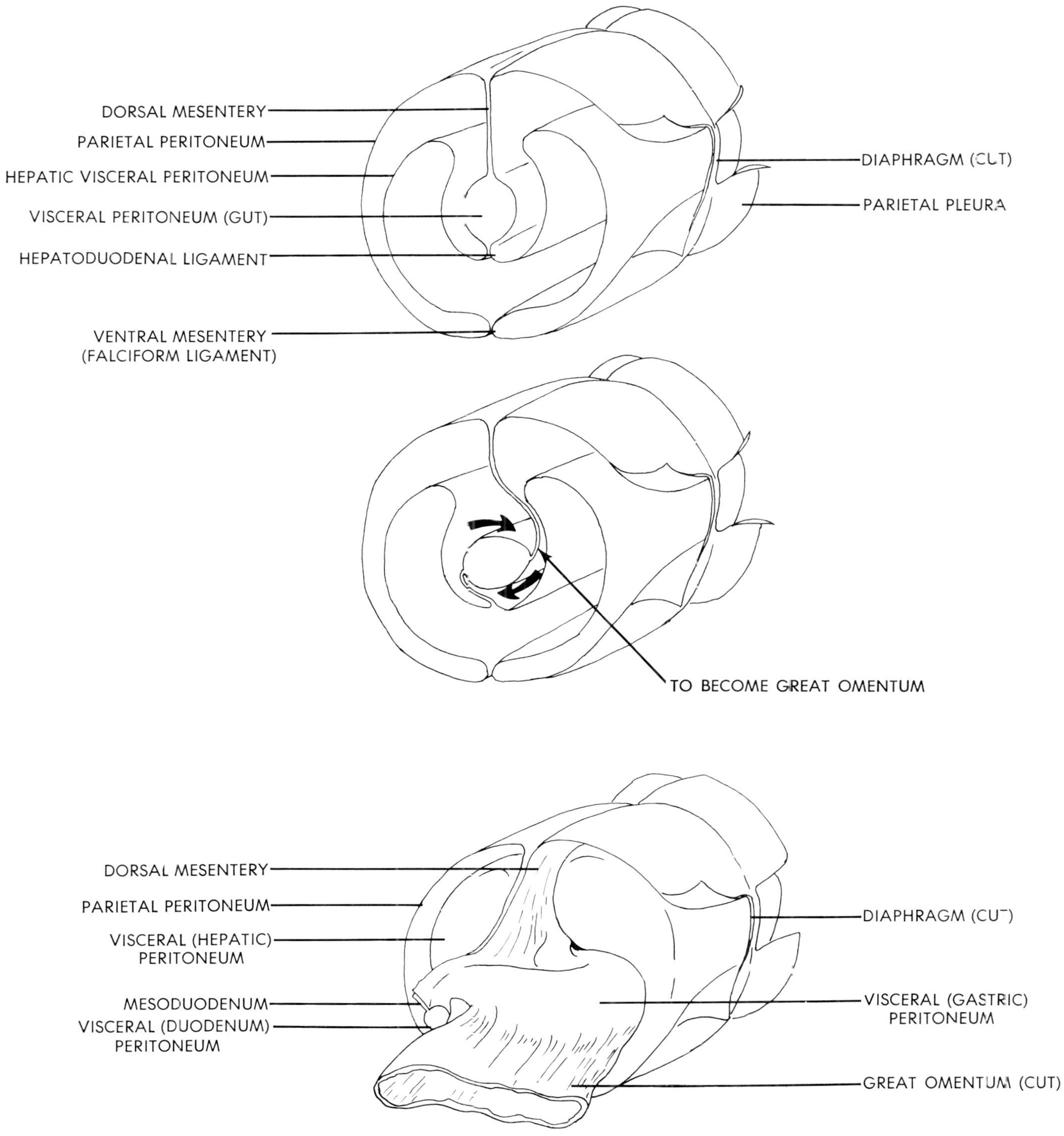

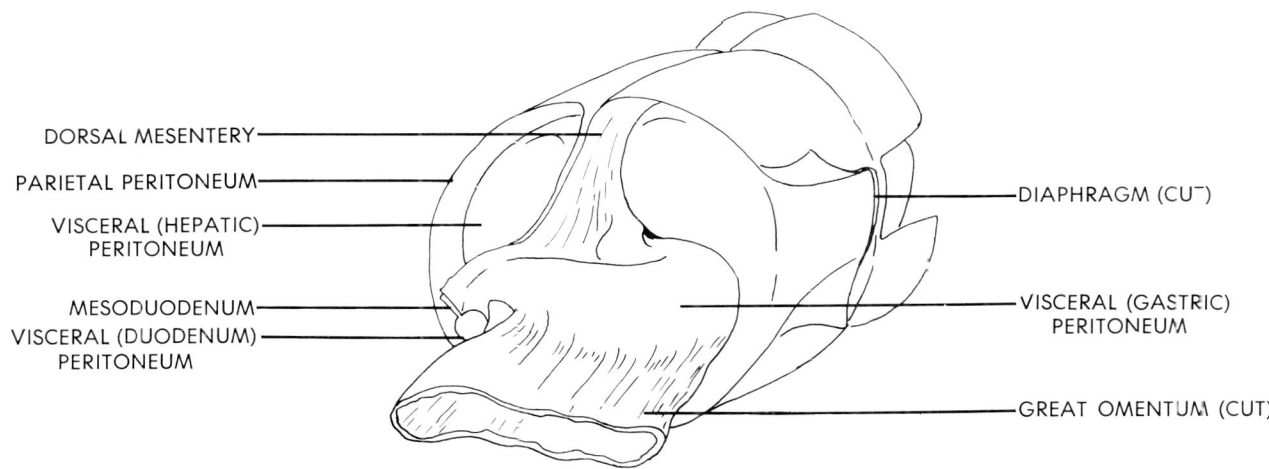

FIGURE 39. Stereodiagrams illustrating stages in the development of some of the abdominal membranes. Contortions of the embryonic gut in the development of the stomach produces the double folded omentum.

suspensory ligament from the ventral body wall to the urinary bladder.

The dorsal peritoneal mesentery is named either according to the visceral structure or portion of the alimentary tract to which it is attached or to describe its structure or appearance. The *mesogastrum* attaches to the stomach (gaster), the *mesoduodenum* to the duodenum, the *mesojejunum* to the jejunum (a portion of the small intestine), the *mesileum* to the ileum (terminal portion of the small intestine), the *mesocecum* to the cecum, and the mesocolon to the large intestine (colon).

Mesenteries attached to the reproductive tissues include the *ligamentum latum uteri* in the female. This in turn may be subdivided into a *mesometrium*, from the uterus proper; a *mesosalpinx*, attached to the oviduct; and a *mesovarium*, suspending the ovary. The *ligamentum latum uteri* is sometimes termed the "broad ligament" because it is very thick.

The name *omentum* is from an appropriate Latin word meaning "fat skin."

The stomach of the cat undergoes some unusual growth movements during embryonic development and this affects the appearance of its mesentery and produces the omentum (see fig. 39). From the entrance of the esophagus the stomach curves to the right side of the abdominal cavity and rotates so the dorsal and ventral surfaces are reversed. The original dorsal half of the stomach (to which the mesentery is attached) undergoes a more rapid growth than does the ventral portion thus producing the large saclike fundic portion. These movements also produce an elongated fold of the mesentery so this membrane comes to lie like a large folded apron (the *greater omentum*) over the ventral surface of the intestines. The parietal attachment of the mesentery (omentum) remains on the dorsal abdominal midline.

THORACIC VISCERA

The thoracic cavity is nearly filled with the lungs and heart. Cranial to the heart and between the lungs locate a *thymus* "gland" covering the large blood vessels and the bifurcation of the *trachea* into primary *bronchi*. If the thymus of your specimen is large, you may not be able to see the bronchi at this time.

ABDOMINAL VISCERA

1. Esophagus. From the pharynx the esophagus passes on the ventral side of the neck, through the thoracic cavity dorsal to the trachea to join the stomach at the extreme cranial end of the abdominal cavity.

2. Liver is the large, dark-brown organ abutting against the diaphragm.

3. Gall bladder, is a dark-green organ imbedded within the right median lobe of the liver.

4. Stomach, a roughly J-shaped structure at the caudal surface of the liver. It is divisible into three regions: the *cardiac* portion at the entrance of the esophagus; the *fundic*, the large bulging middle portion of the stomach; and the *pyloric* region, the narrow caudal portion that joins the small intestine (duodenum). The *greater curvature* is the caudal border of the stomach to which the greater omentum attaches and the *lesser curvature* is the cranial border with the attached lesser omentum.

5. Spleen is an elongated, flat, red organ lying to the left of the greater curvature of the stomach in the greater omentum. The portion of the greater omentum between the stomach and spleen is called the *lienogastric ligament*.

6. Small intestine. The small intestine is divisible into a *duodenum* continuous with the stomach, a *jejunum* between the duodenum and the third portion, the *ileum*. The small intestine is attached to the dorsal body wall by mesentery containing blood vessels and nerves.

7. Pancreas is a two-lobed gland in the mesentery and part of the great omentum between the stomach and the loop of the duodenum.

8. Cecum is a blind pouch at the juncture of the small and large (colon) intestines.

9. Large intestine (colon) may be separated into an ascending colon (on the right side), a descending colon on the left side, and a *rectum*, a straight tube ending at the *anus*. An ill-defined transverse colon is between the ascending and descending portions.

10. Anal glands are paired scent glands opening to the rectum very close to the anus.

11. Kidneys are walnut-sized structures on the dorsal abdominal wall. The right kidney abuts against the liver. These organs are not suspended by mesenteries, but their abdominal surfaces are covered by parietal peritoneum.

12. Adrenal glands are located just cranial and medial to each kidney.

SUGGESTED READING

Allison, J. E., Hoffman, H. H., Faulkner, K. K., and Page, C. H. 1961. Enteric plexus fibers in the mesentery of cats, dogs, and rabbits. *Jour. Comp. Neurol.* 117:383-385.

Griffin, L. E. 1945. A case of incomplete diaphragm with displacement of organs in a cat. *Turtox News*, 23(8): 133.

Chapter 5
Digestive System

The digestive system of the cat begins at the mouth and terminates at the anus. While passing through this system, food is sectioned and subdivided physically by the teeth and jaws; chemically reduced by digestive enzymes; transported through the tract by contractions of the muscular wall; and absorbed through the mucosal wall of the intestine. After absorption the substances are carried back to the liver (a gland of the digestive tract) for further treatment or storage.

Before dissecting further examine the salivary glands exposed when you removed the skin of the head and neck to examine the musculature.

SALIVARY GLANDS

There are five pairs of salivary glands located in the head. The *parotid gland* lies superficially beneath the skin and just below the ear. The *parotid duct* opens just inside the cheek opposite the last premolar tooth.

The *submaxillary* gland is below the parotid at the angle of the jaw. The *submaxillary duct* opens through a small papilla at the base of the frenulum.

The *sublingual gland* is about one inch long and is continuous with the submaxillary.

The *infraorbital gland*, as the name implies, lies within the orbit. Its duct enters the mouth just behind the molar tooth.

The *molar gland* is very small and is located near the cranial edge of the masseter muscle on the outer surface of the maxilla.

The major function of salivary glands is to provide lubrication for food. Secondarily, salivary glands may also secrete an enzyme (amylase) that digests starch to dextrose. The parotid is composed of predominantly enzyme-producing cells but the other salivary glands have mucus-producing cells. The submaxillary gland is approximately equally comprised of enzyme-producing cells and mucus-producing cells. The remaining salivary glands of the cat are almost exclusively mucus secreting.

ORAL CAVITY

Cut through the muscles and skin at the corner of the mouth, then press down the lower jaws with your fingers. Cut the angle of the mandible with bone cutters. Locate the following structures:

1. Cheeks and lips.
2. Vestibule, the space between the lips and teeth.
3. Teeth: incisors, canines, premolars, and molars (see p. 10, fig. 8).
4. Tongue. The musculature of the tongue is seen in figs. 21 and 41 and the surface of the tongue in fig. 40. Some of the important features of the tongue that you should observe at this time are:
   a. Frenulum, connective tissues and skin holding the tongue to the floor of the mouth.
   b. Filiform papillae, spinelike structures on the upper surface of the tongue.
   c. Fungiform papillae, mushroom-shaped structures on the sides and tip of the tongue.
   d. Vallate papillae, set in cuplike depressions on the upper side of the base of the tongue.
   e. Foliate papillae, lie along the sides of vertical folds about lateral to the vallate papillae.
5. Hard palate, roof of forepart of mouth.
6. Soft palate, hangs from back part of hard palate.
7. Nasopalatine ducts open through the forepart of the hard palate.
8. Musculature of the tongue. The musculature of the hyoid draws the tongue forward (*geniohyoideus*), backward (*occipitohyoideus*), and elevates the base of the tongue (*stylohyoideus*). The extrinsic tongue musculature projects the tongue forward, out of the mouth, or moves the tongue from side to side (*genioglossus* and *styloglossus*). Extrinsic muscles also draw the tongue caudally (*hyoglossus*) during swallowing to pass food into the esophagus.

Intrinsic tongue muscles (*lingualis proprius*) alter the shape of the body of the tongue. The lingualis proprius is composed of four sorts of fiber bundles: *Perpendiculares* and *transversae* are located in the center of the body of the tongue and run in the direction their name implies. *Longitudinalis superficialis* fibers are on the dorsal surface of the tongue just beneath the epithelium; the *longitudinalis profundae* fibers are ventral (see fig. 41).

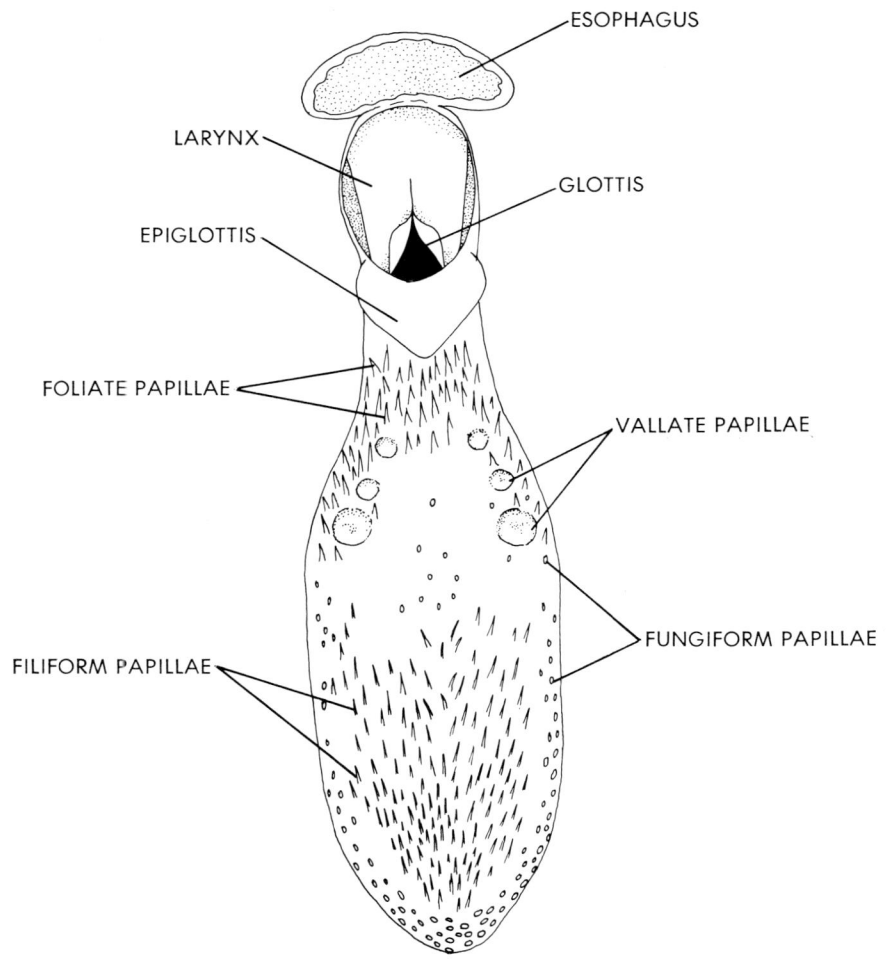

FIGURE 40. Dorsal view of the tongue and pharynx.

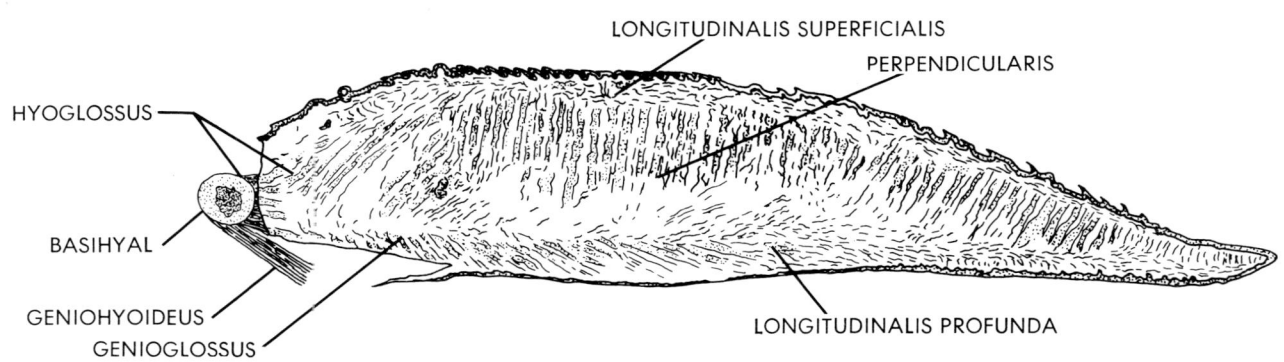

FIGURE 41. *M. lingualis proprius* (intrinsic tongue musculature). Transverse fibers are difficult to see in this section. They run approximately at right angles to the perpendicular fibers.

PHARYNX

The pharynx is the crossing of air and food passages.
1. Fauces, the walls dividing the oral cavity and oral pharynx.
2. Tonsilar fossa, a small pit in the lateral wall at the base of the tongue.
3. Tonsil, a lymphatic nodule inside the tonsilar fossa.
4. Nasopharynx, the cavity above the soft palate (split the soft palate for best observation).
5. Caudal nares, the rostral limit of the nasopharynx.
6. Eustachian tube, an opening in the lateral wall of the nasopharynx to the middle ear cavity.
7. Glottis, the opening into the larynx located just caudal to the epiglottis.

INTESTINAL TRACT AND GLANDS

The wall of the intestinal tract consists of an innermost layer, the *mucosa* that lines the gut cavity (lumen). There is usually a thin layer of muscle—the *muscularis mucosa*—just beneath the mucosa. The *submucosa* is a layer of connective tissue, blood vessels, and nerves extending to the muscularis mucosa. Surrounding all these tissues is a *muscularis externa* consisting of an inner circular and an outer longitudinal layer of smooth muscle. When suspended in the coelom the muscularis externa is enclosed in a serosal membrane (= visceral peritoneum).

The various regions of the intestinal tract of the cat may be identified by the appearance of the mucosa lining that portion of the gut. These distinctions will be made at the appropriate place in the following descriptions.

ESOPHAGUS

The esophagus of the cat is a straight tube that conveys food from the pharynx to the stomach. The mucosa is composed of flat (squamous) cells without a cornified layer. The mucosa—a sparse muscularis mucosa and submucosa—are arranged in longitudinal folds throughout the length of the esophagus. A band of mucous glands occurs in the deeper portion of the submucosa. Though characteristic of mammals, these glands are especially dense in the cat in the upper end (pharyngoesophageal) of the esophagus. The muscularis externa is composed of striated muscle except for a short section of the esophagus near the stomach. There is no serosal membrane around the esophagus, but an external layer of connective tissues, blood vessels, and nerves may be described as a *tunica adventitia*.

STOMACH

The mucosa of the mammalian stomach is composed of tall (columnar) cells as contrasted with the flat cells of the esophageal mucosa. The parts of the stomach may also be distinguished on a histological basis. The divisions of the cat stomach are—*cardiac, fundic,* and *pyloric* regions.

The cardiac stomach is a small area near the esophagus with deep branching, mucus-secreting cardiac glands in the submucosa. The glands of the fundic stomach are deep gastric pits lined with cuboidal (square) cells. Cells lining the upper half of the pits are mucus-producing *chief* cells and those in the lower half secrete *zymogen. Parietal* cells are situated internal to the chief cells (in the submucosa) and secrete a precursor of hydrochloric acid between the chief cells into the lumen of the gastric pits. The chief and parietal cells occur only in the fundic stomach.

The pyloric stomach has coiled, branched glands deep in the submucosa. All of the cells are mucus-secreting and resemble the mucous neck cells of the upper half of the fundic gastric pits.

THE INTESTINES

Villi are present throughout the small intestine but are absent in the large intestine of the cat. The small intestine is divided into three regions—a short *duodenum* adjacent to the pyloric stomach, a long *jejunum* making up half of the remaining small intestine, and an equally long *ileum*. These three regions are best identified by histological features. The duodenum is characterized by the presence of mucus-secreting *Brunner's glands* in its submucosa; these are absent in the remaining small intestine. Aggregates of lymphatic nodules called *Peyer's patches* occur in the wall of the ileum. With some experience Peyer's patches may even be distinguished in a gross dissection. In addition to the presence or absence of Brunner's glands and Peyer's patches the duodenal villi are longer and more densely packed than those in the jejunum or ileum.

As with the small intestine, the large intestine may also be divided into three sections—a blind sack (*cecum*) is caudal to the entrance of the ileum, a long loop with ascending and descending limbs, the *colon*, extends forward from the cecum toward the stomach then turns caudally to join the straight terminal portion of the large intestine, the *rectum*.

The entrance to the large intestine from the ileum is guarded by a sphincter valve, the *ileocolic sphincter*. Absorption of water and some vitamins occurs in the large intestine, but this requires much less absorptive

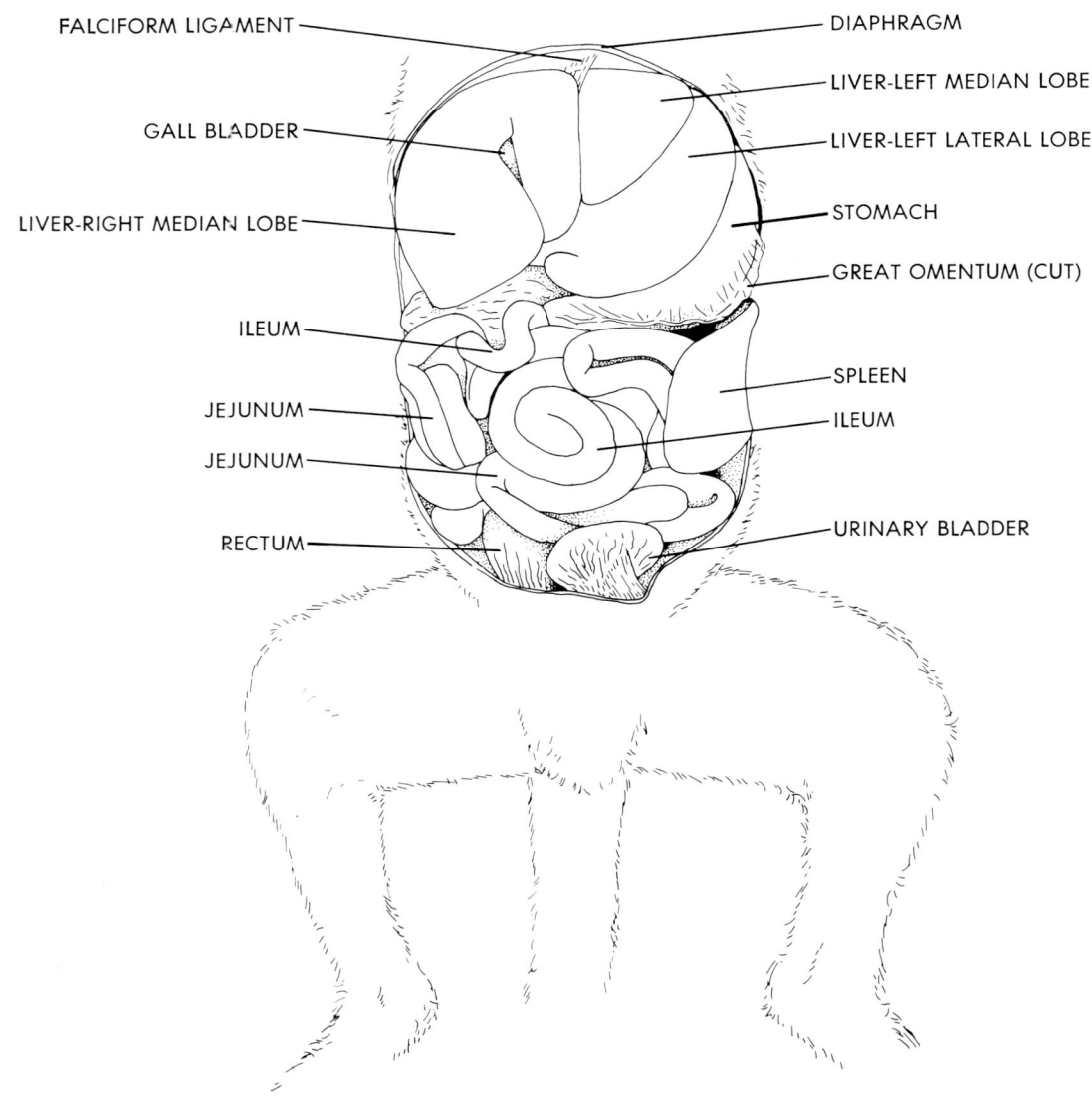

FIGURE 42. Ventral view of the abdominal viscera.

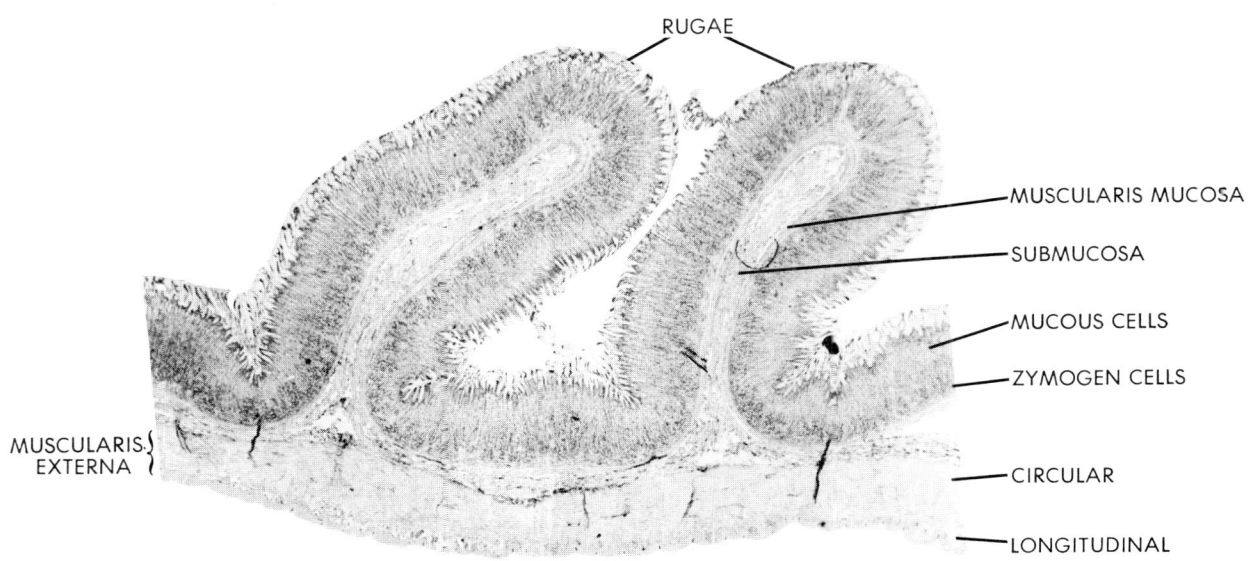

FIGURE 43. Photomicrograph of a cross section of the cat stomach. The columnar cells of the mucosa are arranged in deep pits. Cells at the base of the pits secrete gastric enzymes (zymogen) and thus stain darker than the more superficial mucus secreting cells. The mucosa and submucosa of the cat stomach are arranged in longitundinal folds (*rugae*).

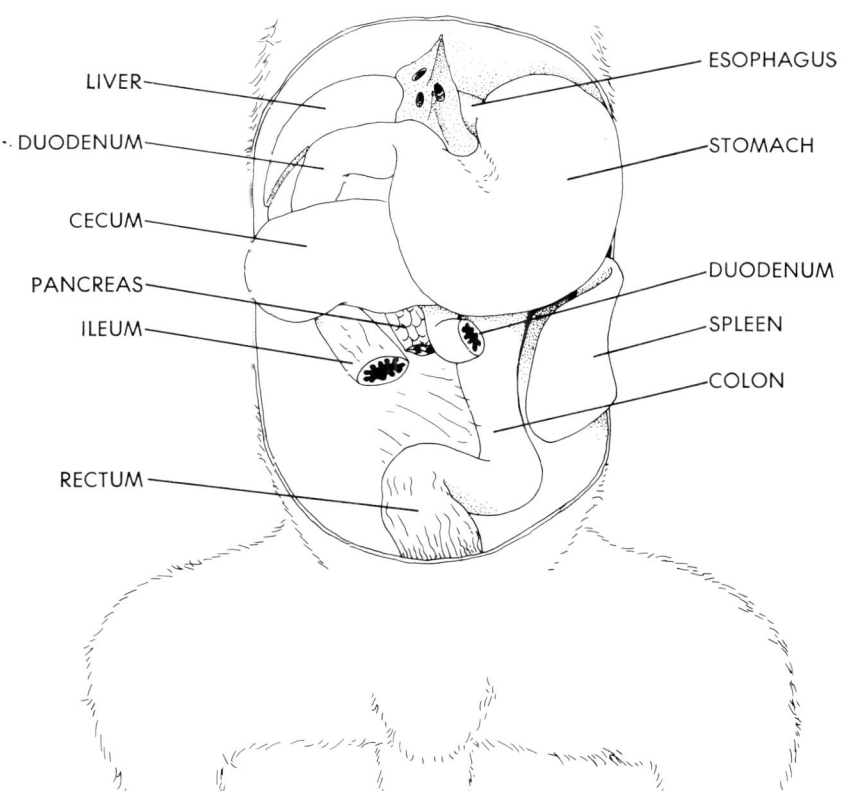

FIGURE 44. Ventral view of the abdominal viscera with the small intestine and omentum removed.

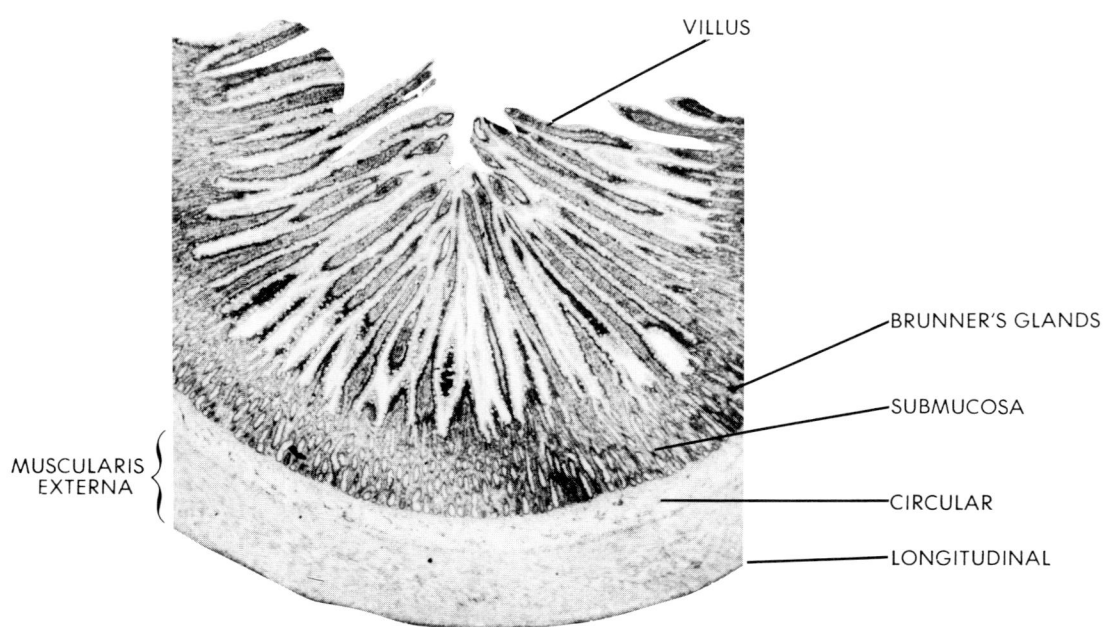

FIGURE 45. Photomicrograph of a cross section of the cat duodenum. The duodenum differs from the rest of the small intestine in having large, branched, mucus-secreting Brunner's glands in the submucosa.

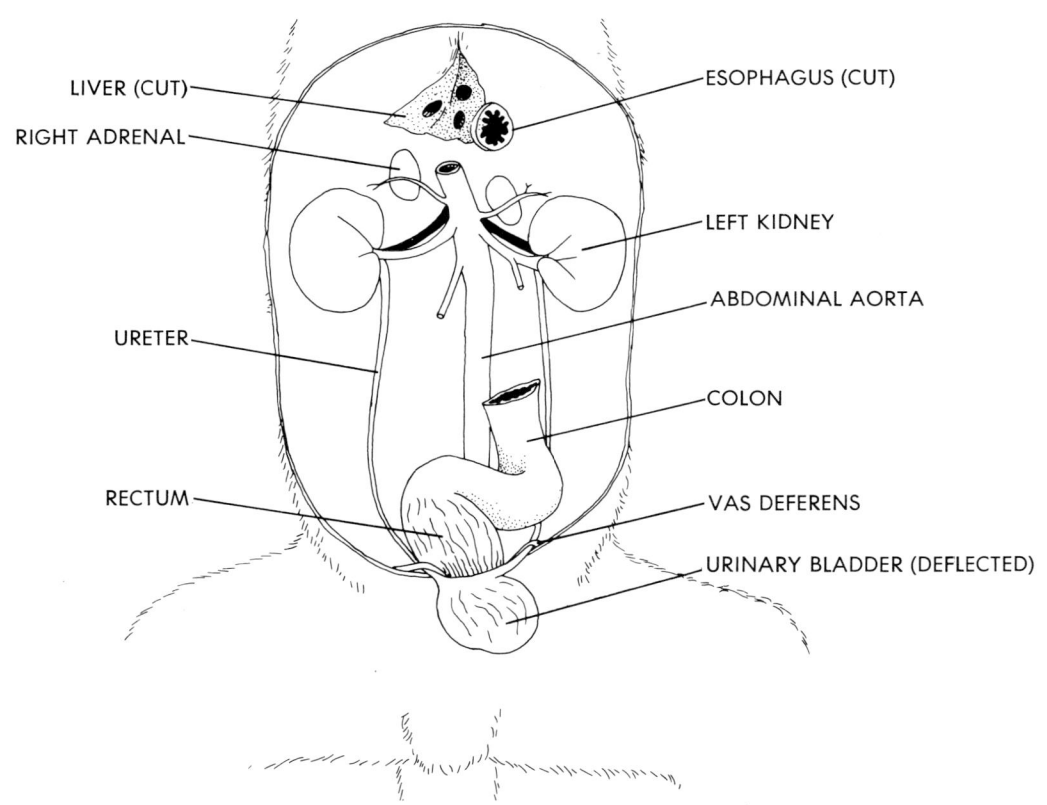

FIGURE 46. Ventral view of the abdominal cavity with most of the viscera removed.

area than that of the ileum. The reduction in absorptive area is evident in the shorter length of the large intestine and the absence of villi.

With the absorption of water the viscosity of the feces increases and must be lubricated for expulsion. A heavy population of mucous cells in the mucosa of the colon and rectum provides the needed lubrication to expel waste material.

Waste material accumulates in the large intestine because the wall of the colon is much less motile than the wall of the small intestine, and the greater diameter of the large intestine allows for the accumulation.

THE DIGESTIVE GLANDS

LIVER

The liver is the largest gland in the cat. The falciform ligament divides the liver into right and left halves. The left half of the liver consists of two lobes: a rostral *left median lobe*, and a caudal *left lateral lobe*. The right half has three lobes: the *right median lobe* is so large it nearly obscures the other lobes of the right half in a ventral view (fig. 42). The right lateral lobe is mainly dorsal to the median lobe and together with the median lobe encloses the *Spigelian* or *caudate* lobe.

The gall bladder is in a cleft on the ventral surface of the right median lobe.

Bile is drained from the liver cells by very small *canaliculi* that run between the individual liver cells. The canaliculi are then collected into *bile ductules*, which in turn drain into the *bile duct* or gall bladder. From the bile duct the bile eventually flows into the duodenum through the *ampulla of Vater*, which is controlled by the *sphincter of Oddi*.

Blood flows into the liver from the hepatic portal vein and the hepatic artery. These larger vessels divide into sinusoids, which collect into central veins. These in turn open to hepatic veins. Lymphatic vessels drain caudally from the liver toward the lesser omentum and duodenum where they join other lymphatics to form a *cisterna chyli* (see fig. 56).

In addition to bile production the liver is also responsible for the conversions of sugar, amino acids, and fats, which are absorbed from the intestines.

The liver of the cat is arranged in lobules. Each lobule is a rough hexagon in cross section with a large central vein near the middle of the lobule. At several of the hexagonal angles are typical hepatic "triads." Each "triad" has a branch of the hepatic portal vein, a branch of the hepatic artery, a ductule collecting bile from several canaliculi.

PANCREAS

The pancreas of the cat is a double-lobed gland. The dorsal lobe is in the duodenal mesentery and is usually drained by two ducts. One duct joins the duct from the ventral lobe, this common duct joins the bile duct from the gall bladder to form the *common bile duct*, which in turn empties into the duodenum. The central duct of the dorsal pancreas may have an additional caudal duct (the duct of *Santorini*) entering the duodenum near the jejunum.

The digestive enzymes of the pancreas are produced in pyramidal-shaped cells arranged in a cluster around a central cavity (acini), which open directly into the pancreatic duct or have a small branch of the duct (called an intercalated duct) projected into the acinar cavity. The acinar cells produce seven or eight different digestive enzymes.

In contrast, bundles or islands of lighter staining cells are interspersed among the acinar groups and produce the hormones, *insulin* and *glucagon*. The lighter staining groups of cells have no central cavity and are known as *isles of Langerhans*.

SUGGESTED READING

Bosma, J. F. 1956. Myology of the pharynx of cat, dog, and monkey with interpretation of the mechanism of swallowing. *Annals Otol.* 65:981-992.

Elbert, M. E. 1956. On the problem of the cyto-architecture of Auerbach's plexus of the small intestine in the cat and dog. *Trans. Mosk. Vet. Akad.* 18:35-38.

Chapter 6
Respiratory System

THORACIC CAVITY
(Figure 47)

Cut through the cartilage on both sides of the sternum. Leave a strip 1/2 inch wide in the center including the sternum and some visceral structures. The thoracic cavity may be separated into three parts: the *pericardial cavity,* into which the *heart* projects; the *right pleural cavity,* into which the *right lung* projects; and the *left pleural cavity,* into which the *left lung* projects. Make another incision at right angles to the above between the two caudal ribs. Bend back the thoracic walls and break or cut the ribs near the spinal column. This will expose the pleural cavity and the lungs.

The *pleura* is the lining of the thoracic coelomic cavity and the lung. The *parietal pleura* lines the cavity, while the *visceral pleura* covers the surface of the lungs.

Pharynx is the passageway common to openings from the nasal chamber (nasopharynx), mouth, esophagus, and trachea (larynx).

Caudal nares are the oral openings to the nasopharynx from the common nasal meatus (above the soft palate).

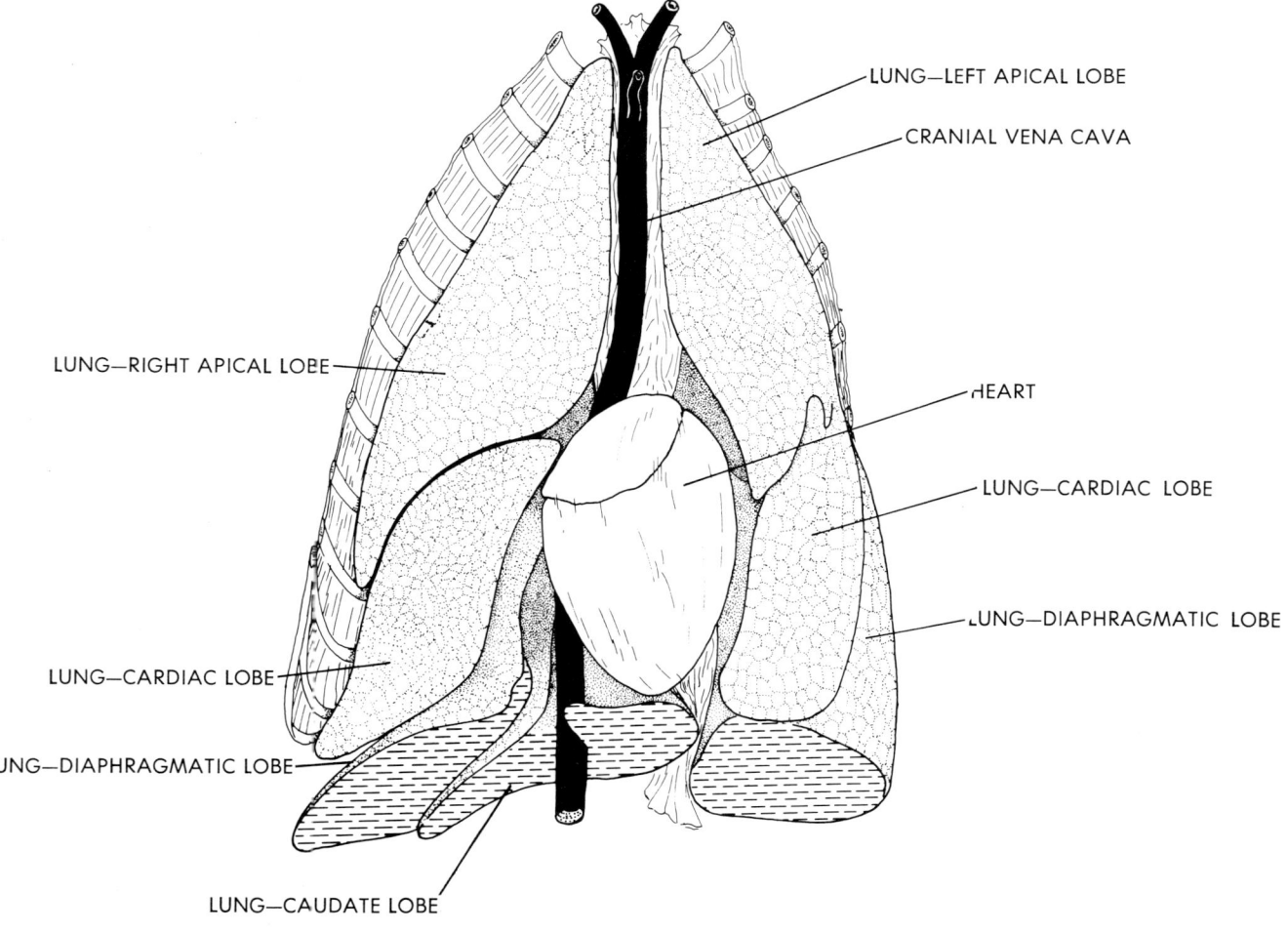

FIGURE 47. Ventral view of the thoracic viscera.

Glottis is the name of the opening into the larynx from the pharynx.

Larynx is the "voice box." The wall of the larynx is held rigid by five cartilages; a rostral *epiglottis* projects into the pharynx to the glottis. The caudal border of the glottis is supported by small paired *arytenoid* cartilages. The center of the larynx is stiffened by a single *thyroid* (shield-shaped) cartilage having paired articulations with the hyoid apparatus. The thyroid cartilage is not a complete cartilaginous ring as is the more caudal *cricoid* cartilage. The vocal cords are attached to the arytenoid and epiglottal cartilages.

Trachea (fig. 21), or windpipe, is held open by a series of U-shaped cartilages in the wall of the trachea. The lining of the trachea is composed of ciliated, mucus-secreting cells. The trachea bifurcates into the *bronchi* at the level of the heart just after it enters the thoracic cavity.

Bronchi branch from the trachea on each side of the thoracic cavity—one bronchus going to each lung. The bronchial tubes are also supported by U-shaped cartilages and lined by ciliated cells. Within the lungs, the bronchi subdivide into *secondary* and *tertiary bronchi* with cartilage "rings," but the smaller branches or *bronchioles* have no skeletal supports. The bronchioles ultimately terminate in small chambers termed *vestibules,* which are, in turn, bordered by *alveoli,* somewhat like tiny grapes in a cluster.

Lungs are the large, paired, lobed structures occupying most of the thoracic cavity. The right lung is composed of four lobes while the left lung has only three divisions. Between the two lungs is a space, the *mediastinum,* surrounded by visceral plurae and containing several visceral organs (see p. 60).

SUGGESTED READING

Hartroft, W. S. 1947. Comparative pulmonic alveolar size in man, cat, rabbit, monkey, guinea pig, goat, dog, baboon, rat and mouse. *Anat. Record* 97:417.

Kastrinos, Wm. 1964. Lung comparison. *Turtox News* 42 (1):18-19.

Niewenhuis, R. 1966. Comparative histology of the trachea of the dog and cat. *Anat. Record* 154:454.

Rosengren, J. H. 1966. The cat's meow. *Turtox News* 44 (6):154-155.

Sobin, S. S., Fung, Y. C., Tremor, H. M., and Rosenquist, T. H. 1972. Elasticity of the pulmonary alveolar microvascular sheet in the cat. *Circulation Research* 30:440-450.

Sobin, S. S., Tremor, H. M., and Fung, Y. C. 1970. Morphometric basis of the sheet-flow concept of the pulmonary alveolar microcirculation in the cat. *Circulation Research* 26:397-414.

Chapter 7
Circulatory System

The mammalian circulatory system consists of the heart, arteries, veins, capillaries, and lymphatic vessels. Because of their smallness, the lymphatic vessels and capillaries are difficult to locate and your instructor may wish to omit them. To study the arteries and veins, the vessels should be injected with colored latex. Triply injected specimens will have an *hepatic portal system* injected with yellow latex. Doubly injected specimens will have arteries filled with red latex, and veins, except for the hepatic portal system, filled with blue latex.

If any of the injection mass reaches the pulmonary vessels, these will have a reverse color scheme; that is, the veins will be red and the arteries blue.

As you dissect, carefully clean the connective tissues away from the vessels. Do not use a scalpel; forceps and a blunt probe will be adequate for this dissection.

Now is really the time when you will wish you had selected a skinny specimen for dissection. Fat is deposited around blood vessels, and clearing away adipose tissue during your dissection is a tedious and messy job. Study the figures of vessels carefully, then hold a pair of forceps, gently but firmly, over the exposed portion of the vessels to be cleaned. Next, carefully pull the forceps over the vessel, thus separating the fat from the blood vessel. After the chunks of fat are removed, wipe the vessel and adjacent areas with a paper towel to remove the grease released when you broke the fat cells with your forceps.

THE HEART (Figure 48)

Remove the ventral body wall over the heart at this time. Find the *pericardial sac,* then open the sac to expose the heart and blood vessels. The *pericardium* is made up of two membranes. One membrane is the *parietal pericardium,* which is similar to the parietal pleura and parietal peritoneum in its formation. The *visceral pericardium* is not attached to the parietal pericardium by a mesentery. The "pericardial mesentery" disappeared during the very early embryonic development of the heart. The second membrane of the pericardium is a portion of the mediastinal region of the parietal pleura. Locate and study the following structures:

1. Left ventricle, the chamber of the caudal end of the apex of the heart.
2. Right ventricle, the chamber slightly cranial to the left ventricle.
3. Right atrium, the chamber cranial to the right ventricle.
4. Left atrium, the chamber cranial to the left ventricle.
5. Auricule, a small muscular lobe from each atrium that extends caudally over part of the ventricle.
6. Pulmonary arch, on the cranial part of the right ventricle as it passes between the auricular appendages. The pulmonary arch branches into right and left pulmonary arteries which further subdivide before reaching the lungs.
7. Aorta, a large arch beginning at the left ventricle. Notice the branches leading in various directions from the aorta. It may be necessary to remove some fat before these blood vessels can be examined.
8. *Caudal* and *cranial vena cavae* enter the right atrium. *Pulmonary* veins enter the left atrium.

The heart should now be examined in more detail. Cut the pulmonary arch, the caval veins, the pulmonary veins, and the aortic arch close to the heart and remove the heart. Note the stub of aortic arch remaining with the heart. With a pair of fine tip forceps loosen and then pull out the latex. This latex will be a negative impression of the semilunar valve of the aorta and will probably include the beginnings of the coronary arteries. Make an incision through the wall of each atrium and cut off (transversely) the apex of the ventricles. Wash and pick out (with forceps) the clotted blood. Locate the following internal structures:

9. Coronary sinus enters the right atrium through a small valve near the entrance of the caudal vena cava. The coronary sinus receives blood from the musculature of the heart.
10. Tricuspid valve prevents blood in the right ventricle from entering the right atrium. There are said

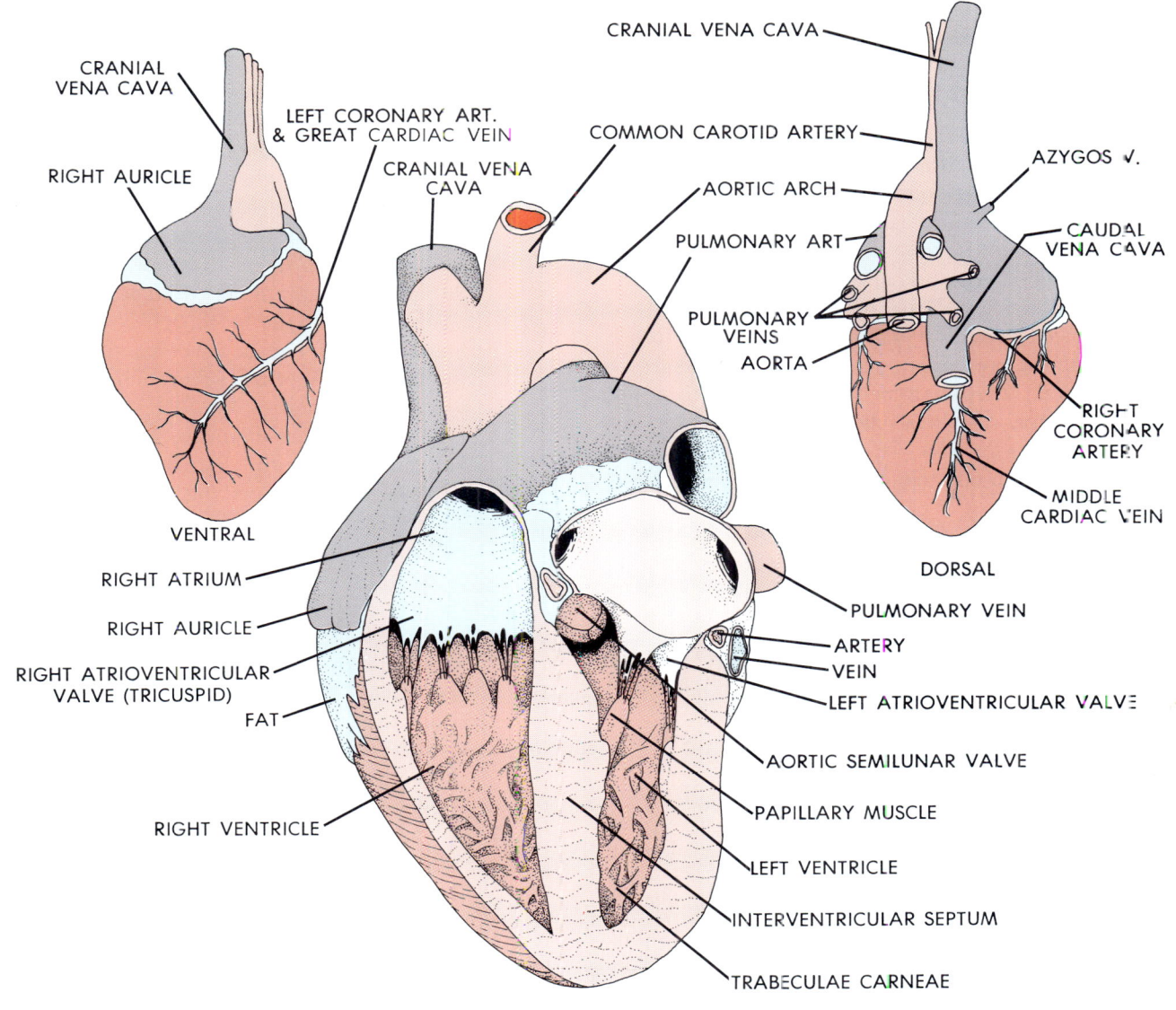

FIGURE 48. The heart of the cat. Upper left figure is a ventral view and the upper right is a dorsal view. The center figure is a view from the left side of the heart with most of the left wall dissected away to show internal structures. The right atrium is behind the right atrioventricular valve.

to be three flaps to this valve, but these will be very difficult to distinguish in the cat heart. The flaps are held to the walls of the right ventricle by *chordae tendineae*. The chordae tendineae attach to raised muscle bundles, the *papillary muscles*.

11. Trabeculae are muscle bundles or strands lining the ventricles. Papillary muscles are "raised" trabeculae.
12. Pulmonary semilunar valve, between the pulmonary arch and the right ventricle. Note the three flaps of this valve from the ventricular side. Compare this view with the latex image of the aortic semilunar valve.
13. Aortic semilunar valve, between the aortic arch and the left ventricle. See discussion after no. 8 p. 72.
14. Bicuspid valve is similar to the tricuspid (see no. 10 above) valve, except that it has two flaps rather than three. This is sometimes called the *mitral*

valve because of its supposed resemblance to a bishop's mitre (cap).

VESSELS CRANIAL TO THE HEART

A. Arteries cranial to the Heart (Figures 48, 50 and 51).
 1. Innominate artery, largest artery branching from the aortic arch. Its three major branches are: right subclavian, and left and right common carotids.
 2. Common carotid arteries branch from the branchiocephalic trunk and pass forward on the either side of the neck just lateral to the trachea.
 3. Caudal thyroid artery is the first branch of the common carotid at the base of the neck and it passes forward on the lateral walls of the trachea to anastomose with the cranial thyroid artery.
 4. Cranial thyroid artery is the second branch of the common carotid at the level of the thyroid gland. A muscular artery branches at the same level and passes to the back of the neck.
 5. Internal carotid artery is a small branch at the level of the occipital and enters the skull at the carotid canal medial to the bulla.
 6. External carotid artery is the large continuation of the common carotid to the head. This vessel will appear to be the same as the common carotid and the internal carotid appears as a branch of the major trunk. Several small branches leave the external carotid just after the branching of the internal carotid. These include a combined occipital and ascending pharyngeal trunk and branches to the salivary glands.
 7. Lingual artery is a large branch from the external carotid which serves the tongue and floor of the mouth. The external carotid passes on the outer surface of the auditory bulla to the eye orbit and forms a rete plexus at the origin of the ocular rectus muscles with branches that join the circle of Willis with the internal carotid.
 8. External maxillary artery, to the upper and lower jaws, and to the lips from the external carotid at the angle of the jaws near the branch of the lingual artery.
 9. Internal maxillary artery continues from the external carotid rete within the orbit to the palate and through the infraorbital canal to the rostrum.
 10. Right common carotid artery. The branches from this are similar to those of the left common carotid.
 11. Right subclavian artery is a continuation of the innominate.
 12. Vertebral artery arises from the subclavian at about the region where that artery leaves the thoracic cavity. It supplies neck muscles by small branches, then passes to the intervertebral foramen of the sixth cervical vertebra. At the base of the brain the two fuse and form the basilar artery to the brain.
 13. Sternal artery (or internal mammary), from the ventral side of the subclavian, supplies the ventral body wall.
 14. Costocervical axis branches from the subclavian just opposite the first rib. It later branches into two arteries supplying the back and neck.
 15. Thyrocervical axis arises from the subclavian below the first rib. Its branches supply the neck and shoulder.
 16. Axillary artery, the continuation of the subclavian to the brachium. The ventral thoracic artery and the long thoracic artery branch from the axillary and serve the chest muscles. The subscapular also branches from the axillary and divides into several branches; the most important is the *thoracodorsal* artery that supplies the latissimus dorsi and other nearby muscles.
 17. Brachial artery is the continuation of the axillary artery beyond the point of branching of the subscapular. It gives off several branches to the foreleg muscles above the elbow.
 18. Radial artery is the continuation of the brachial artery below the elbow. It sends branches to the lower arm and to the toes.
 19. Left subclavian artery gives off branches similar to those of the right subclavian. It should be traced to fix the arteries firmly in mind.
 20. Intercostal arteries branch from the thoracic aorta and supply the intercostal muscles.
 21. Esophageal arteries branch from the thoracic aorta and supply the esophagus.
 22. Cranial phrenic arteries serve the diaphragm.

B. Veins Cranial to the Heart (figs. 49, 50, 51 and 54).
 Very little dissection will be required to expose the veins since they usually accompany the arteries. Begin at the heart and proceed toward the head. Locate the following:
 1. Cranial vena cava joins the right atrium. The next four veins drain into this large vein.

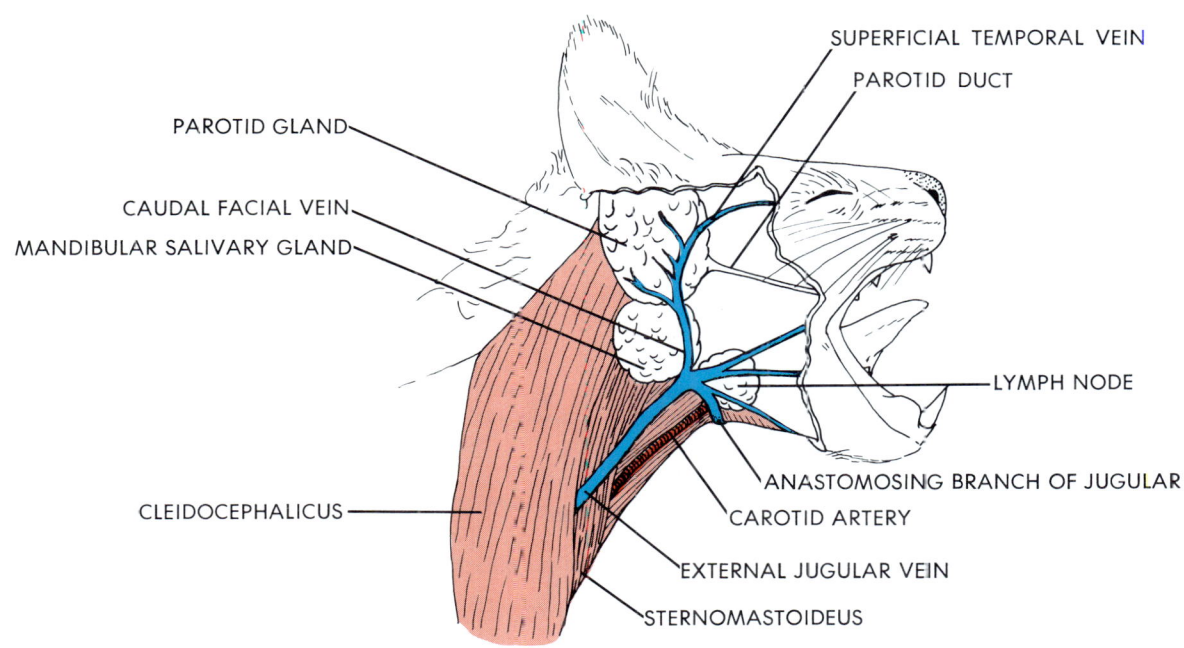

FIGURE 49. Superficial veins of the head and neck.

2. Azygos vein, just cranial to the point of entrance of the cranial vena cava, into the heart. The *intercostal veins* from between the ribs, the *bronchial veins* from the bronchi, and the *esophageal veins* from the esophagus all drain into the azygos vein.
3. Internal mammary veins, a pair, but united at the root to form the sternal vein.
4. Vertebral vein, usually unites with the costocervical before entering the cranial vena cava.
5. Brachiocephalic (or innominate) veins unite at the cranial margin of the thoracic cavity to form the cranial vena cava.
6. Costocervical vein joins the vertebral vein just prior to the entrance into the brachiocephalic. It receives branches from the back and neck muscles.
7. External jugular vein drains the head and face and joins the brachiocephalic together with the *subclavian vein*.
8. Subclavian vein receives the axillary vein from the shoulder.
9. Axillary vein receives the brachial vein from the arm and the subscapular vein.
10. Brachial vein parallels the brachial artery.
11. Subscapular vein drains into the axillary vein from the upper arm and shoulder.
12. Ventral thoracic vein enters the axillary lateral to the subscapular. It drains the pectoral muscles.
13. Long thoracic vein joins the axillary just lateral to the ventral thoracic. It also comes from the pectoral muscles.
14. Thoracodorsal vein joins the axillary near the long thoracic vein after draining the latissimus dorsi muscle.
15. Cranial facial extends from the face, jaws, and submaxillary gland to the external jugular.
16. Caudal facial receives the *cranial* and *caudal auricular* and the *superficial temporal* veins before joining the external jugular.
17. Transverse jugular unites the right and left external jugulars at the throat.
18. External jugular forms by the union of the facial veins and empties into the subclavian (see nos. 7 and 8 *above*).
19. Cephalic vein enters the external jugular at the base of the neck. The cephalic vein receives the superficial veins of the arm. There is no arterial circulation corresponding to this.

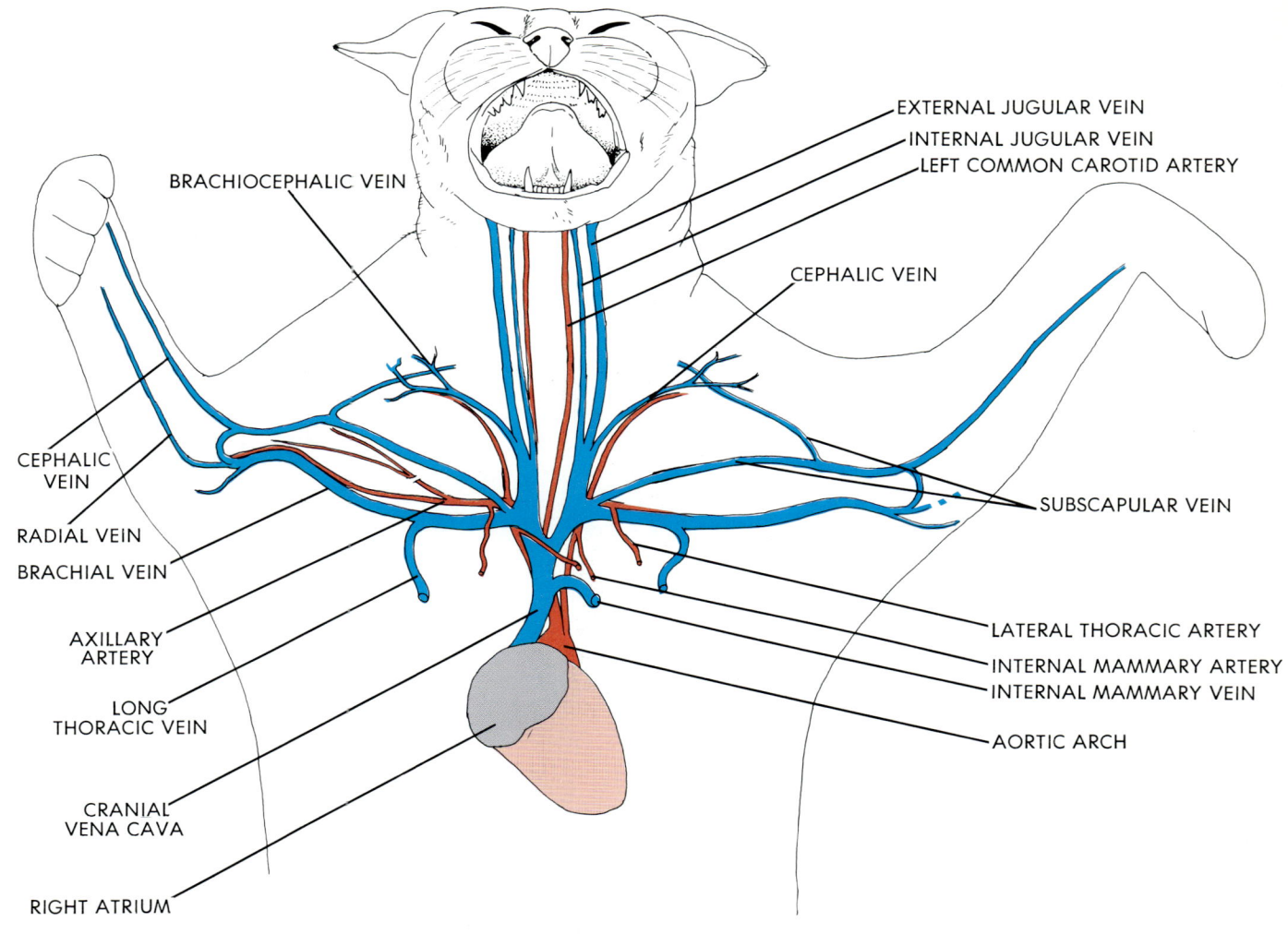

FIGURE 50. Major vessels of the neck.

20. Internal jugular vein joins the external jugular at the base of the neck, and the two jugulars (internal and external) enter the subclavian by a common trunk. The internal jugular drains the brain, passes through the jugular foramen and receives additional vessels from the caudal facial region and the vertebral column.

VESSELS CAUDAL TO THE HEART

A. Arteries of the Abdominal Region (figs. 52, 53).
   Trace the thoracic aorta from the thoracic cavity through the diaphragm. It is now known as the abdominal aorta.
   1. Celiac artery extends from the aorta just caudal to the diaphragm. The phrenic arteries sometimes branch from the celiac artery to the diaphragm, but they usually come from the adrenolumbar arteries.

   a. Splenic artery branches from the celiac and supplies the spleen, pancreas, the greater curvature of the stomach, and the omentum.
   b. Left gastric artery supplies the lesser curvature of the stomach.
   c. Hepatic artery supplies the liver. In the liver this vessel fuses with the hepatic portal vein. The hepatic gives rise to the *gastroduodenalis* before entering the liver.
   d. Gastroduodenalis divides into three arteries: *pyloric, cranial pancreaticoduodenal* and *right gastroepiploic*. The pyloric fuses with the left gastric artery; the cranial pancreaticoduodenal fuses with the caudal pancreaticoduodenal (of the cranial mesenteric); and the right gastroepiploic fuses with the splenic.

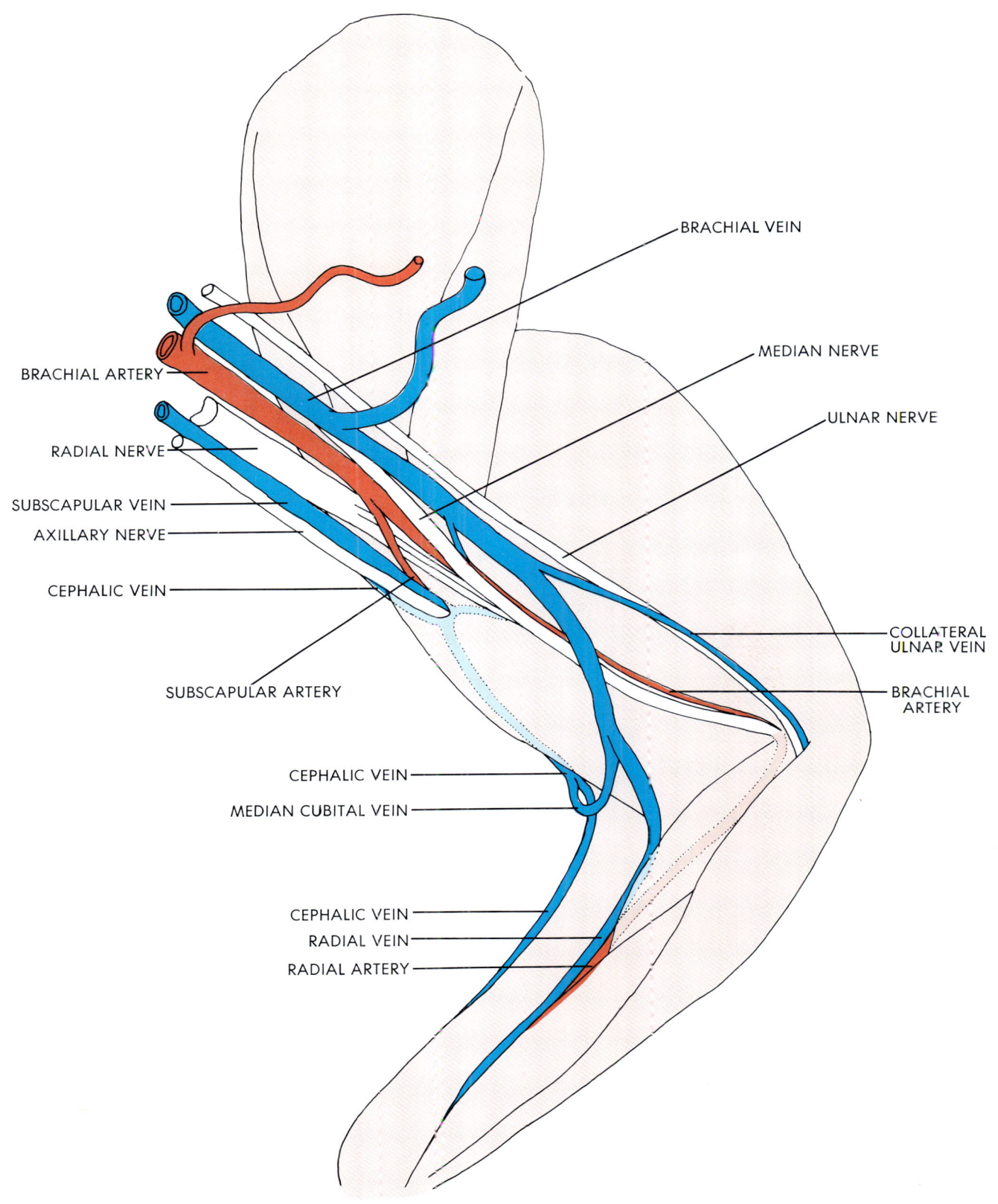

FIGURE 51. Arteries, veins, and nerves of the axilla and brachium.

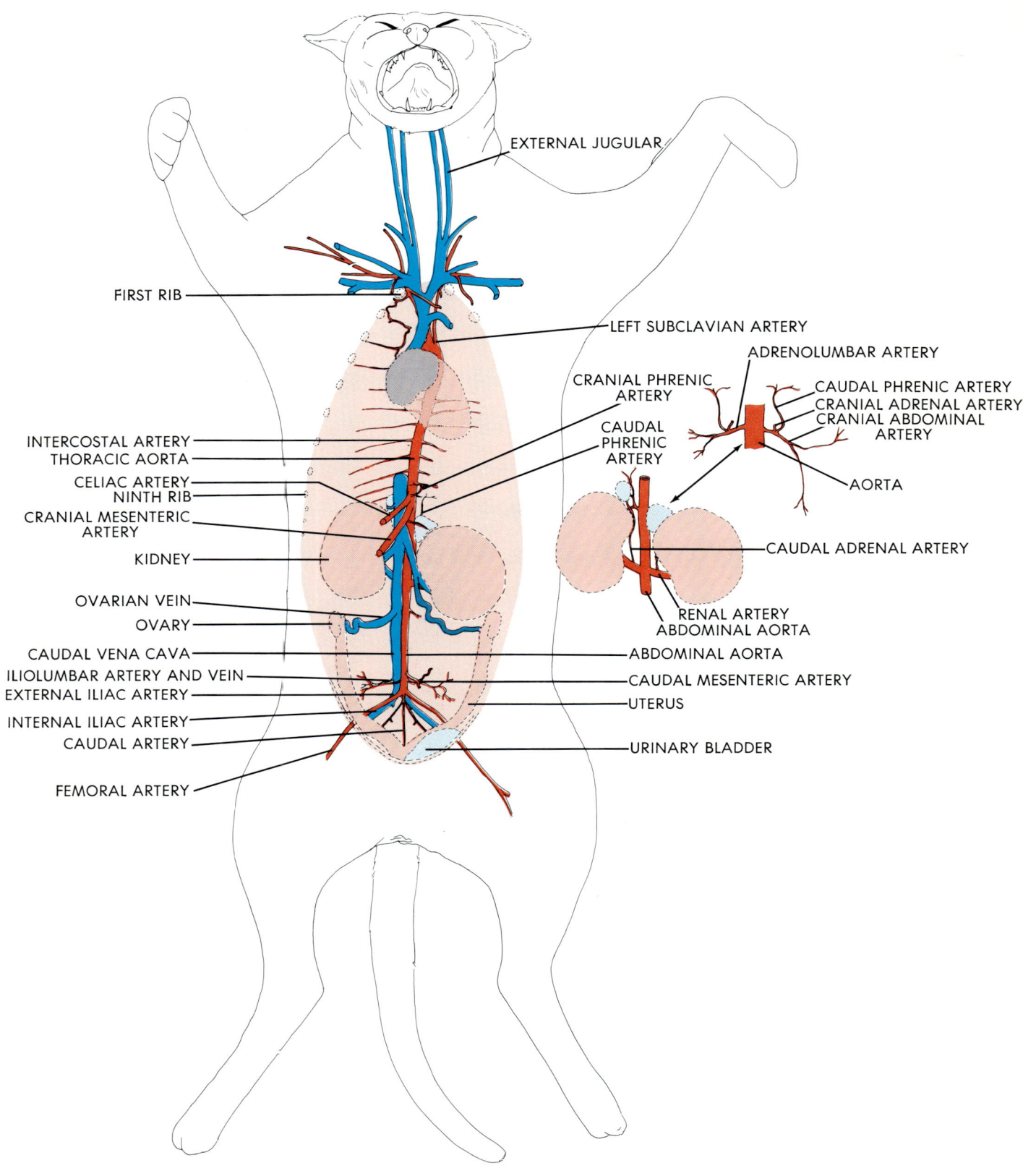

FIGURE 52. Major arteries of the trunk.

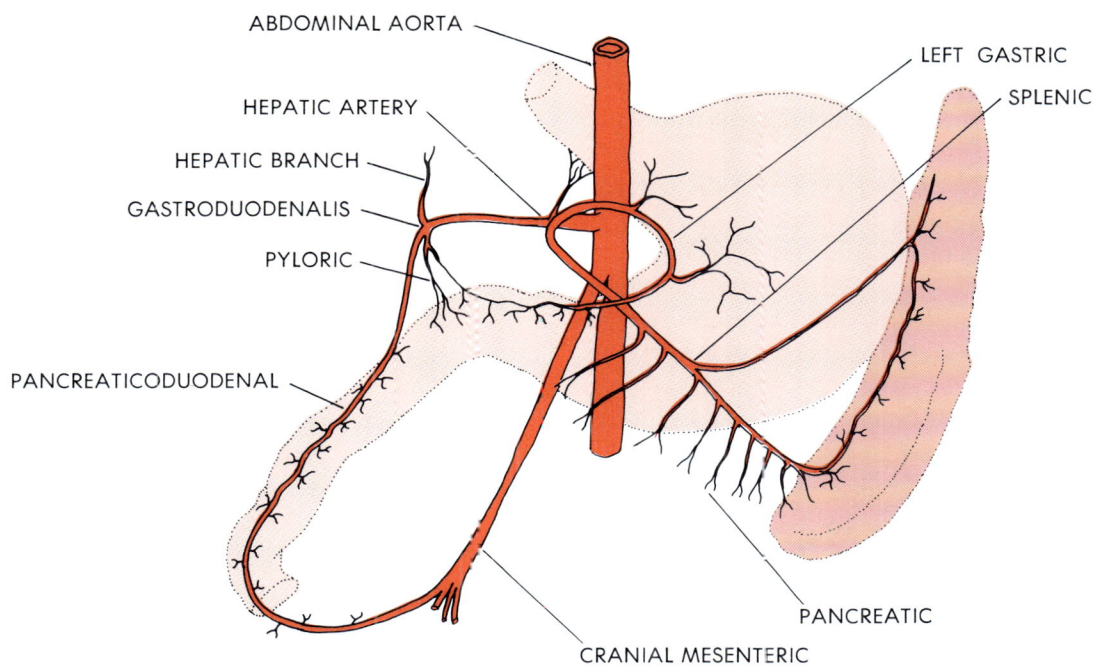

FIGURE 53. The celiac axis.

2. Cranial mesenteric artery branches from the aorta and supplies the intestines and the pancreas.
   a. Middle colic artery branches from the cranial mesenteric and supplies the colon.
   b. Caudal pancreaticoduodenal artery branches from the cranial mesenteric and supplies the duodenum and pancreas.
   c. Right colic artery branches from the cranial mesenteric and supplies the ascending colon.
   d. Iliocolic artery branches from the cranial mesenteric and supplies the cecum and the lower ileum.
   e. Intestinal arteries branch from the cranial mesenteric and supply the intestines.
3. Adrenolumbar arteries arise from the aorta just caudal to the cranial mesenteric artery and supply the adrenal glands and the back.
   Caudal phrenic artery branches from each adrenolumbar artery to the diaphragm (see no. 1 above).
4. Renal arteries branch from the aorta to the kidneys.
5. Gonadal arteries, ovarian (in the female) and spermatic (in the male), branch from the aorta. The ovarian arteries supply the ovaries, and the spermatic arteries supply the testes. The left gonadal vessel usually branches from the left renal artery.
6. Lumbar arteries branch from the aorta to the back muscles.
7. Caudal mesenteric artery, large unpaired branch from the aorta caudal to the genital arteries, supplies the lower colon and the rectum.
   a. Left colic artery branches from the caudal mesenteric and supplies the colon.
   b. Superior hemorrhoidal artery branches from the caudal mesenteric and supplies the rectum.
8. Iliolumbar arteries branch from the aorta caudal to the caudal mesenteric and supply the iliopsoas muscles.
9. External iliac arteries (a pair of them) branch from the aorta at the caudal end of the abdominal cavity. Be careful not to injure the urogenital organs while looking for these arteries. They subdivide into two branches, which supply the hind legs.
   a. Deep femoral artery branches from the external iliac. It branches into the caudal *epigastric artery* to the abdomen, and into other branches to the genital organs and the bladder.

b. Femoral artery branches from the external iliac and supplies the leg. A branch in the upper thigh is the *cranial femoral artery.*
10. Internal iliac arteries branch caudal to the external iliacs.
    a. Umbilical artery branches from the internal iliac and supplies the bladder.
    b. Middle hemorrhoidal artery branches from the internal iliac caudal to the umbilical and supplies the rectum. The *uterine artery* branches from the middle hemorrhoidal in the female.
11. Caudal artery (or sacral artery), the continuation of the aorta caudal to the iliacs; it supplies the sacrum and the tail.

B. Veins of the Abdominal Cavity (figs. 53, 54).

Very little dissection will be needed here, but be careful not to disturb the urogenital organs and the nerves. Locate the following:

1. Caudal vena cava enters the right atrium.
2. Phrenic veins enter the caudal vena cava from the diaphragm.
3. Hepatic veins, entering the caudal vena cava from the liver, can be seen by dissecting out the liver around the caudal vena cava.
4. Right adrenolumbar vein enters the caudal vena cava just caudal to the liver. It drains the right adrenal gland and part of the back.
5. Right renal vein, from the kidneys, enters the caudal vena cava just caudal to the right adrenolumbar.
6. Left adrenolumbar enters the caudal vena cava slightly caudal to the right renal; drains the left adrenal gland.
7. Left renal vein enters the caudal vena cava slightly caudal to the left adrenolumbar; drains the left kidney. (The left adrenolumbar may join the left renal vein before it enters the caudal vena cava. The *left gonadal vein* of cats enters the left renal.)
8. Right internal spermatic vein enters the caudal vena cava slightly caudal of the kidneys.
9. Right ovarian vein enters the caudal vena cava just caudal to the kidney.
10. Lumbar veins enter the caudal vena cava at varying intervals. Lift the caudal vena cava carefully to find these entering on the dorsal side.
11. Iliolumbar veins enter the caudal vena cava from the groin region.
12. Common iliac veins unite just caudal to the iliolumbar to form the caudal vena cava.
13. Caudal vein runs from the tail to the right common iliac.
14. External iliac vein: the common iliac is formed by the union of this vessel with the *internal iliac.*
15. Internal iliac vein joins the external iliac to form the common iliac.
16. Femoral vein, from the leg, joins with the deep femoral to form the external iliac vein.
17. Deep femoral vein joins with the femoral to form the external iliac just caudal to the entrance of the internal iliac.
18. Middle hemorrhoidal vein joins the internal iliac and drains the rectum and bladder.
19. Gluteal veins enter the internal iliac and drain the hip region.

C. Hepatic Portal System (fig. 55).

This is a group of veins that begin and end in a set of capillaries. The liver capillaries converge into hepatic veins (see no. B.3 *above*) from the liver to the caudal vena cava. The hepatic portal system takes blood from the digestive organs of the abdomen and the spleen to the liver. Find the hepatic portal vein by turning the lobes of the liver until the bile duct is exposed. The portal vein is in the hepatoduodenal ligament near the bile duct. Locate the following branches of the hepatic portal vein:

1. Coronary vein, extends from the lesser curvature of the stomach to the hepatic portal.
2. Pancreaticoduodenal vein empties to the hepatic portal slightly caudal to the coronary and drains the cranial part of the duodenum and the pancreas.
3. Right gastroepiploic vein, runs from the greater curvature and pyloric region of the stomach and joins the hepatic portal vein.
4. Gastrosplenic vein joins the cranial mesenteric to form the portal vein just caudal to the gastroepiploic. It is formed by the *right* and *left splenic veins* and a *pancreatic vein.*
   a. Left splenic vein comes from the omentum, greater curvature of the stomach, and the spleen.
   b. Middle gastroepiploic veins join the gastrosplenic just prior to the latter's entrance into the portal vein. These veins come from the pyloric region of the stomach.
   c. Pancreatic vein joins the gastrosplenic just prior to the latter's entrance into the portal. This vein drains the pancreas.

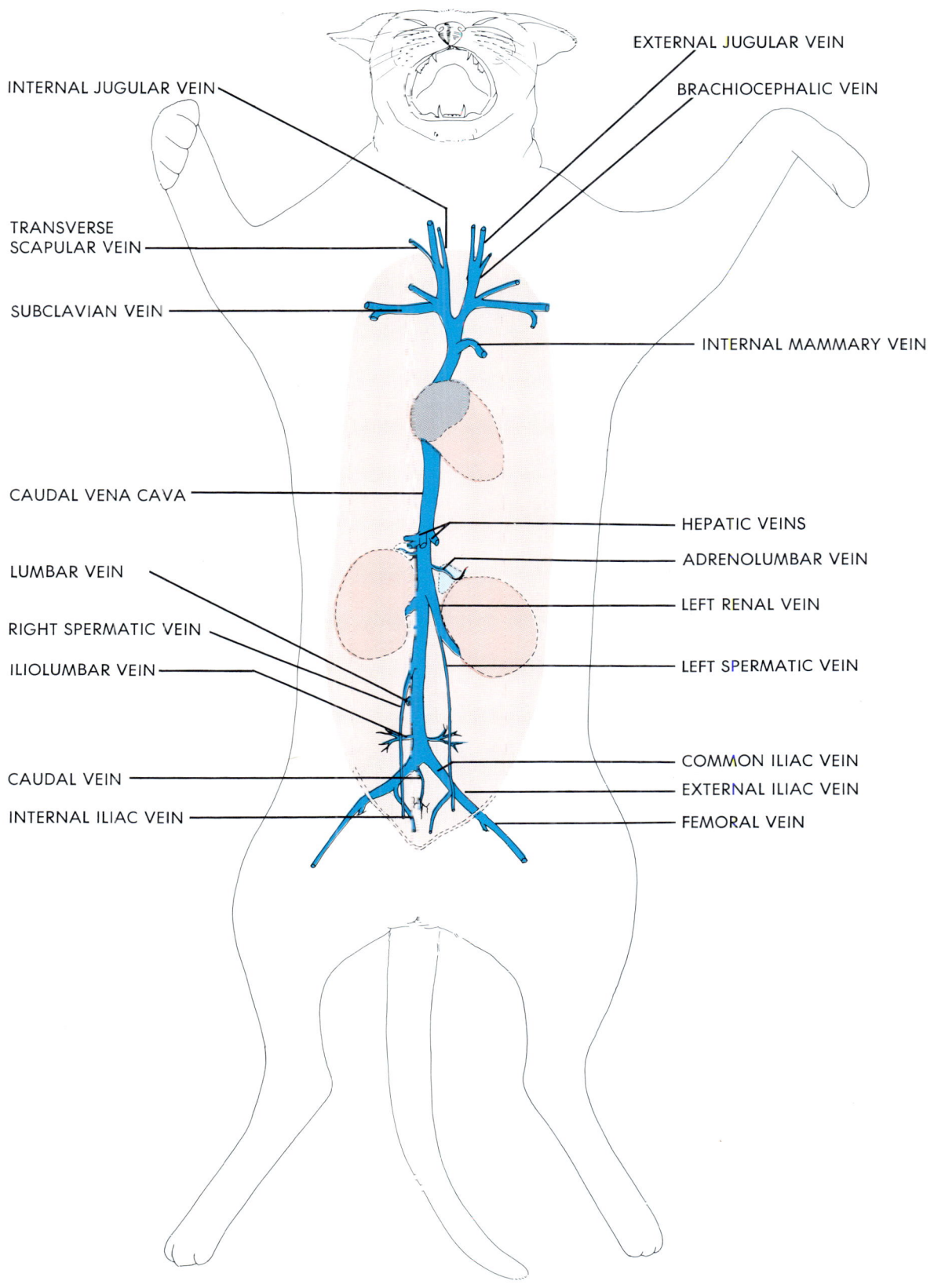

FIGURE 54. Major veins of the cat.

5. Cranial mesenteric vein is formed by the union of several intestinal veins, the caudal pancreaticoduodenal and the caudal mesenteric.
   a. Caudal pancreaticoduodenal vein enters the cranial mesenteric near the caudal border of the pancreas and duodenum.
   b. Caudal mesenteric vein joins the cranial mesenteric just caudal to the junction of the cranial mesenteric and caudal pancreaticoduodenal veins. It drains the rectum and colon.
   c. Intestinal veins join the cranial mesenteric caudal to the junction of the caudal mesenteric. These veins drain the small intestine.

THE LYMPHATIC SYSTEM

Lymphatic vessels are an extension of the veins and carry the nearly colorless lymph from the lacteals in the villi of the small intestine and the blind capillaries of the lymphatic spaces to the base of the external jugular veins.

In addition to capillaries and lacteals, the lymphatic system also includes vessels, nodes, follicles, spleen, and thymus.

Lymphatic organs function in the formation of certain white blood cells and in filtering of particulate matter. Some lymphatic structures are also responsible

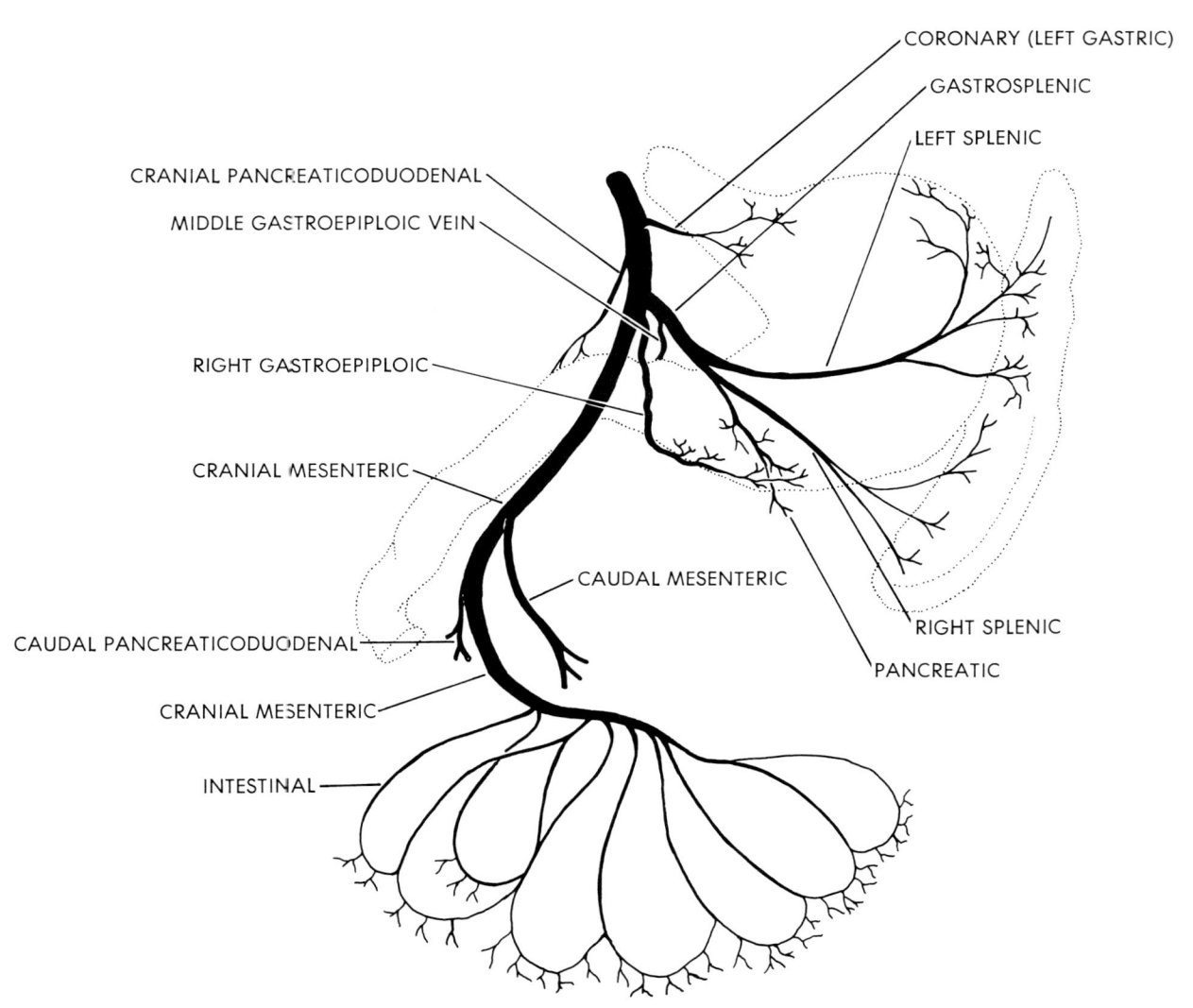

FIGURE 55. The hepatic portal system.

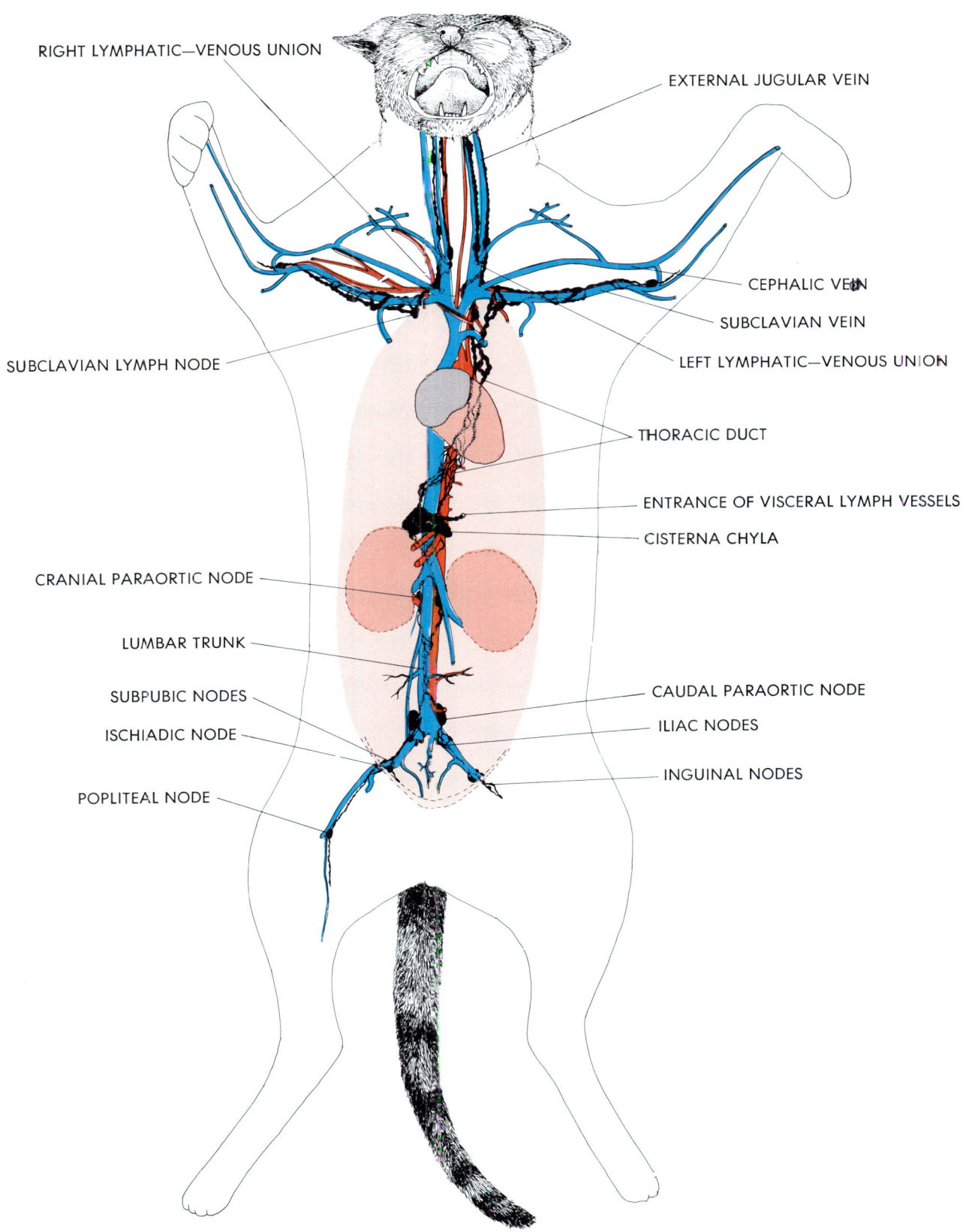

FIGURE 56. Lymphatic vessels of the cat.

for the formation of antibodies, which protect the body from various diseases:

1. *Lymphatic vessels* are very thin-walled tubes that break easily under pressure. The various vessels drain into two major trunks, the *thoracic duct* and the right lymphatic trunk. The entrances of the lymphatic ducts into the jugular veins are sometimes filled with latex from the veins. These injected trunks will probably be the only lymphatics you will be able to see in the cat.

   The thoracic duct originates at the cisterna chyli and extends forward on the dorsal surface of the aorta to the area of the aortic arch. At this point the thoracic duct passes over the left wall of the esophagus to the junction of the external jugular and subclavian veins. The thoracic duct opens to the venous system at the junction of the left subclavian and external jugular, or at the juncture of the left internal and external jugulars, or at both locations. Before entering the jugular vein, the thoracic duct is joined by two jugular lymph trunks from the head and neck and a subclavian lymph trunk from the left arm.

   There is no thoracic duct on the right side, but the two right jugular trunks and the right subclavian trunk usually join together before entering the right jugular vein.

   The deep lymphatic jugular trunks (right and left) and the thoracic duct of the cat have numerous valves that give the vessels a beaded appearance.

2. *Lymph Node* is a nodule of lymphoid tissues with lymph vessels entering and leaving the lymphoid mass. The vessels of the lymph node are equipped with valves directing the flow of lymph through the node and toward the vessels draining into the thoracic ducts. The lymph nodes are scattered throughout the body but are especially prominent in the axillary and inguinal regions, at the base of the neck and at the angle of the jaws, and in the dorsal mesentery of the small intestine.

3. *Lymph follicles* consist of adenoid (glandlike) tissue arranged in discrete clusters similar to nodes. the tonsils (see p. 65) are examples of lymph follicles.

4. *Thymus* is a glandlike bilobed structure on the cranial ventral surface of the heart and, in young animals, extending cranially on the ventral surface of the neck just lateral to the trachea, as far forward as the thyroid gland.

5. *Spleen* is an unpaired, flat, red organ lying on the left side of the peritoneal cavity just caudal to the stomach (see p. 62 and figs. 42, 43). In addition to its role as a lymphatic organ the spleen removes exhausted erythrocytes from circulation and dissociates their hemoglobin into bilirubin (bile pigment) and iron that is restructured into hemoglobin within new erythrocytes.

THE ENDOCRINE ORGANS

The endocrine glands are not really an anatomical organ system but may be grouped together because they all release their various products into the circulatory system. These glands are not described in this manual in a single chapter but they are illustrated (and mentioned) at various points throughout the text as follows:

Pituitary, figs. 61, 62, pp. 92, 93.

Thyroid, lies on the ventral surface of the trachea near the larynx.

Parathyroid (embedded in the thyroid; cannot be seen in a gross dissection).

Pancreas (isles of Langerhans) fig. 44, pp. 67 and 69.

Adrenal, figs. 46, 52, pp. 68 and 78.

Testes, fig. 57, p. 87.

Ovary, fig. 58, p. 88.

SUGGESTED READING

Ackart, R. J., Shaw, J. S., and Lawrence, J. S. 1940. The blood cell picture of normal cats. *Anat. Rec.* 76:357-363.

Anufriew, W. N. 1928. The nerves of the heart of the cat. *Zeitschr. f. Anat. u. Entwg.* 86:639-654.

Ayer, A. A. and Rao, Y. G. 1958. The coronary arterial patterns in some common laboratory animals: rabbit, dog and cat. *Jour. Anat. Soc.*, India 7:5-8.

Biscoe, T. J. and Bucknell, A. 1963. The arterial blood supply to the cat diaphragm with a note on the venous drainage. *Quart. Jour. Exper. Physiol.* 48(1):27-33.

Biscoe, T. J., Lall, A., and Sampson, S. R. 1969. On the nerve endings associated with the carotid body glomus cells of the cat. *Jour. Physiol.*, London 200:131-132.

Bradshaw, P. 1958. Arteries of the spinal cord in the cat. *Jour. Neurol., Neurosurgery and Psychiatry*, 21(4):284-289.

Chungcharoen, D. M., Daly, deB., and Schweitzer, A. 1952. The blood supply of the carotid body in cats, dogs and rabbits. London *Jour. Physiol.* 117:347-358.

Coleridge, H., Coleridge, J.C.G., and Howe, A. 1967. A search for pulmonary arterial chemoreceptors in the cat with a comparison of the blood supply of the aortic bodies in the new-born and adult animal. *Jour. Physiol.*, London 191:353-374.

Davis, D. D. 1941. The arteries of the forearm in carnivores. *Field Mus. Nat. Hist. Zoology* 27:137-227.

Easton, T. W. 1967. Two venous anomalies of commercial laboratory specimens. *Turtox News* 45(2):66-67.

Elias, H. 1945. Comparative histology of domestic animals, III. Endocrine glands. 1. The adrenal gland. *Middlesex Veterinarian*, spring-summer.

Etter, C. L. 1965. An anomalous left precaval vein of the cat. *Turtox News* 43(12):294.

Ghoshal, N. G. 1972. The arteries of the pelvic limb of the cat *(Felis domesticus)*. *Zbl. Vet. Med. A.* 19:78-85.

Groulade, F. 1969. Hematology of the normal cat. *Bull. Acad. Vet.* 42(8):811-814.

Hadziselimovic, H., Secerov, D., and Gmaz-Nikulin, E. 1974. Comparative anatomical investigation on the coronary arteries in wild and domestic animals. *Acta anat.* 90:16-35.

Howe, A. 1956. The vasculature of the aortic bodies in the cat. *Jour. Physiol.*, London 134:311-318.

Huntington, G. C. and McClure, C.F.W. 1920. The development of the veins in the domestic cat. *Anat. Record* 20:1-31.

Hutchison, R. L. 1969. An unusual anomaly. *Carolina Tips* 32(5):17.

Kampmeier, O. F. 1969. *Evolution and Comparative Morphology of the Lymphatic System.* Springfield, Ill.: Charles C. Thomas, Publisher.

Lancaster, L. Y. 1945. An anomaly involving the hepatic portal and systemic systems of the cat. *Turtox News* 23(5):83-84.

Latimer, H. B. 1964. The anomalous right subclavian artery. *Turtox News* 42(8):214-216.

McMullen, D. B. and Clark, W. W. 1938. An anomaly of the venous system in a cat, showing paired superior and inferior vena cavae. *Trans. Ill. State Acad. Sci.* 31:247-248.

McChesney, J. M. and Smith, H. M. 1944. The hepatic portal system of the cat. *Ward's Nat. Sci. Bull.* 18.

McKibben, J. S. and Getty, R. 1968. A comparative morphologic study of the cardiac innervation in domestic animals. II. The Feline. *Amer. Jour. Anat.* 122(3):545-553.

Nader, I. A. and Kamille, S. J. 1963. A common subclavian artery in the cat. *Turtox News* 41(8):198-199.

Schermer, S. 1954. *The Morphology of the Blood of Laboratory Animals: The Cat.* Leipzig: J. A. Barth, pp. 11A-92.

Troll, R. 1967. An anomaly of the median sacral and internal iliac arteries in *Felis domestica*. *Turtox News* 45(4):123-124.

Zucchero, P. T. 1964. An anomaly of the right subclavian artery of the cat. *Turtox News* 42(1):14.

Chapter 8
Urinary and Reproductive Organs

The *kidneys* (figs. 52, 54, 57) will be found embedded in fat on the dorsal body wall. The kidneys are not suspended by a mesentery as are the other abdominal organs and they are covered by peritoneum only on the portion adjacent to the abdominal cavity. Because of this fact the kidneys are said to be *retroperitoneal*. Remove some of the fat around the kidneys but be careful not to destroy the adrenal gland and blood vessels that are also embedded in fat in this region.
Find the following structures:
1. Hilus, the medial concave opening to the kidney, permitting access by the renal blood vessels and ureters.
2. Ureter, the duct extending from the kidney to the urinary bladder. Remove the right kidney from the body and slice it longitudinally into dorsal and ventral halves.
3. Renal pelvis, the expanded portion of the ureter within the renal sinus.
4. Renal sinus, the cavity of the kidney which is entered via the hilus.
5. Renal papilla, the apex of the medulla extending into the renal pelvis.
6. Renal cortex, the outer portion of the kidney distinguishable by color from the medulla.
7. Renal medulla, the inner body of the kidney, darker in color than the cortex.
8. Urinary bladder, a reservoir for urine from the ureter. The bladder is held to the *ventral* body wall by a mesentery, the suspensory ligament.
9. Urethra, the duct extending from the neck of the bladder to the exterior. In the male, this duct extends through the penis and receives the male sex products. In the female this duct opens to the vestibule inside the urogenital orifice.

MALE REPRODUCTIVE ORGANS
(Figures 54, 57)

Locate the following structures:
1. Scrotum, a large sac of skin, muscle, and connective tissue on the exterior of the body just ventral to the anus and containing the testes. Cut open the scrotum and note the following layers from the exterior to the inner wall: (1) integument; (2) cremaster muscle and fascia; (3) tunica vaginalis.
2. Testes, the paired reproductive glands within the tunica vaginalis. Contains the seminiferous tubules.
3. Spermatic cord extends from the testes into the abdominal cavity. It is composed of the ductus deferens, spermatic artery and vein, and nerve, bound together in a sheath.
4. Epididymis, a highly coiled tubule connecting the ductuli efferentes and ductus deferens. The *caput epididymis* covers the cranial end of the testes and leads to the *corpus epididymis,* which is applied to the lateral surface of the testes. The corpus empties to the *cauda epididymis* on the caudal end of the testes. The cauda in turn joins the ductus deferens.
5. Ductuli efferentes are very small tubules from the seminiferous tubules opening to the caput epididymis.
6. Seminiferous tubules, microscopic tubules within the testes where the spermatozoa are formed.
7. Ductus deferens, the continuation of the cauda epididymis to the urethra. Trace one of these (right or left) from its exit from the spermatic cord to the urethra.
8. Prostate glands, small bulb-shaped glands covered by striated muscles and located on either side of the urethra near the entrance of the ductus deferens.
9. Penis consists of three cavernous bodies (two *corpora cavernosa penis* and one *corpus cavernosum urethrae*) a glans (the enlarged distal end), and a baculum or penis bone. Cross section the penis and note the cavernous bodies.
10. Cowper's, or bulbourethral, glands, are situated on either side of the penis at the attachments of the bulbocavernosus and ischiocavernosus muscles to the penis.

FEMALE REPRODUCTIVE SYSTEM
(Figures 52, 58)

Locate the following structures:
1. Membranes include the mesenteries or ligaments

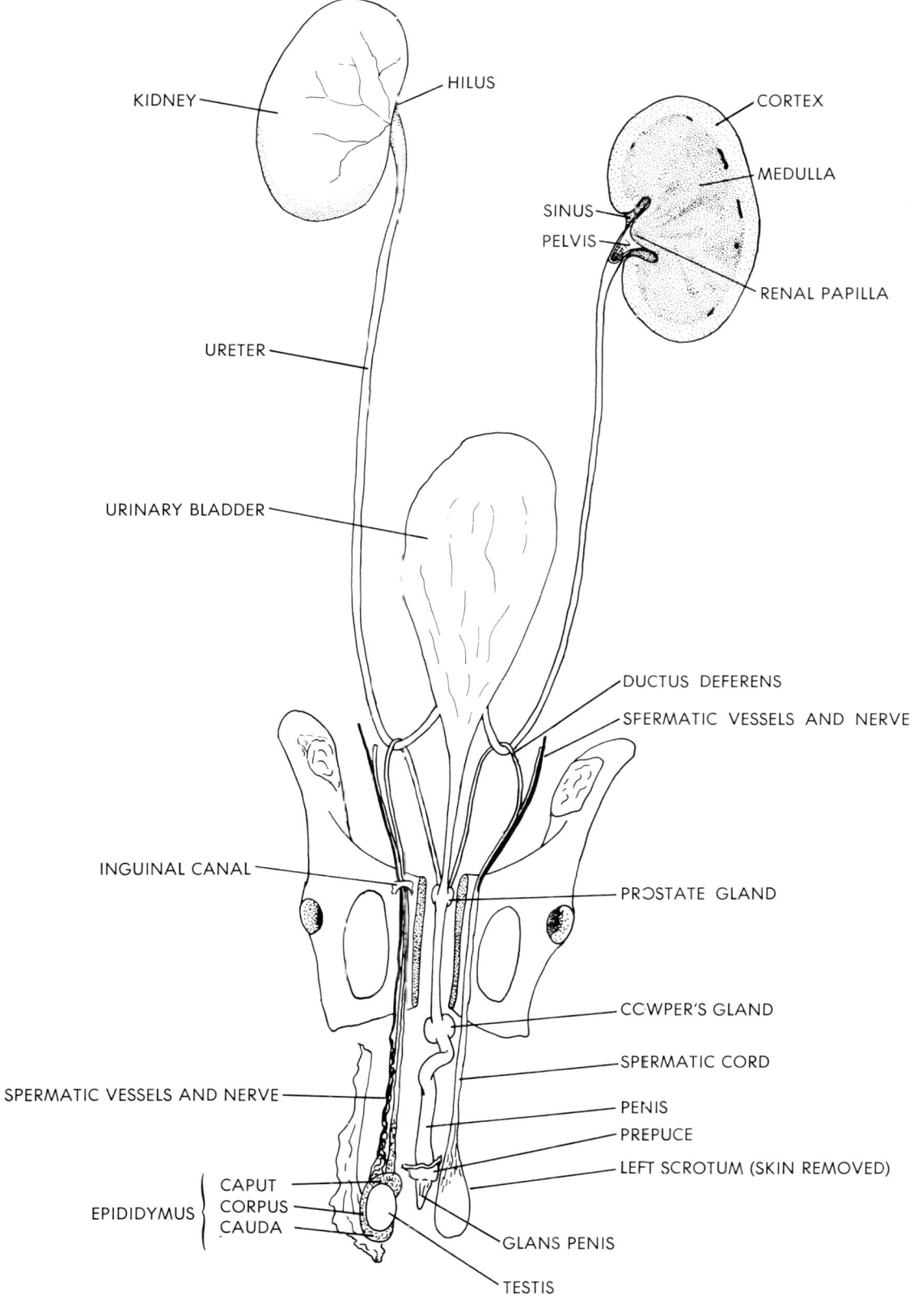

FIGURE 57. The excretory system and male reproductive organs, ventral view.

and peritoneum supporting or covering the genital organs. There are two ligaments (= mesenteries) on each side in the cat. The *broad ligament* supports the ovary, uterine tubes, and uterus to the dorsal body wall. The *round ligament* is a sheet of connective tissue from the horn (= *cornu*) of the uterus to a point on the body wall that corresponds to the inguinal canal in the male. A short *ovarian ligament* holds the ovary to the broad ligament.

2. Ovaries are the female reproductive glands. The eggs develop by meiotic divisions inside fluid-filled chambers called *follicles*. The follicles increase in size as the egg develops until it reaches the mature stage. The mature follicle is known as a *graafian follicle*. After the egg erupts, the walls of the follicle become the *corpus luteum*. Eventually the corpus luteum degenerates into a *corpus albicans* (white body). The corpus luteum produces female sex hormones.

3. Uterine tubes are the convoluted ducts which carry the egg to the uterus. These tubes are sometimes called *oviducts*, *Müllerian ducts*, or *Fallopian tubes*. The internal opening to the uterine tube is a funnel-shaped expansion known as the *ostium*. Fingerlike projections from the free edge of the ostium are called *fimbriae*. The interior of the uterine tubes and ostia are lined with cilia. Movements of the fimbriae and cilia establish a current that first carries the egg into the uterine tube after the follicle erupts and then on into the uterus.

4. Uterus is the chamber in which the embryo develops. In the cat the uterus consists of two horns (*cornu*) joined caudally in a *body*. This type of uterus is called a *bicornuate* or *bipartite* uterus. The human uterus has only the median body without the horns. This type is called *simplex*. Caudally, the muscular terminal end of the body of the uterus projects into the vagina as the *cervix* of the *uterus*.

5. Vagina is the terminal chamber of the female reproductive tract. The vagina opens caudally into the *vestibule* or *urogenital sinus* together with the urethra (from the bladder). In order to examine the vagina and associated structures you will need to cut away the pubic symphysis. Use bone cutters to cut the ischium and pubis on each side. Cut away the pelvic musculature and connective tissue with scissors and remove the central piece.

6. Vestibule is a chamber common to both the repro-

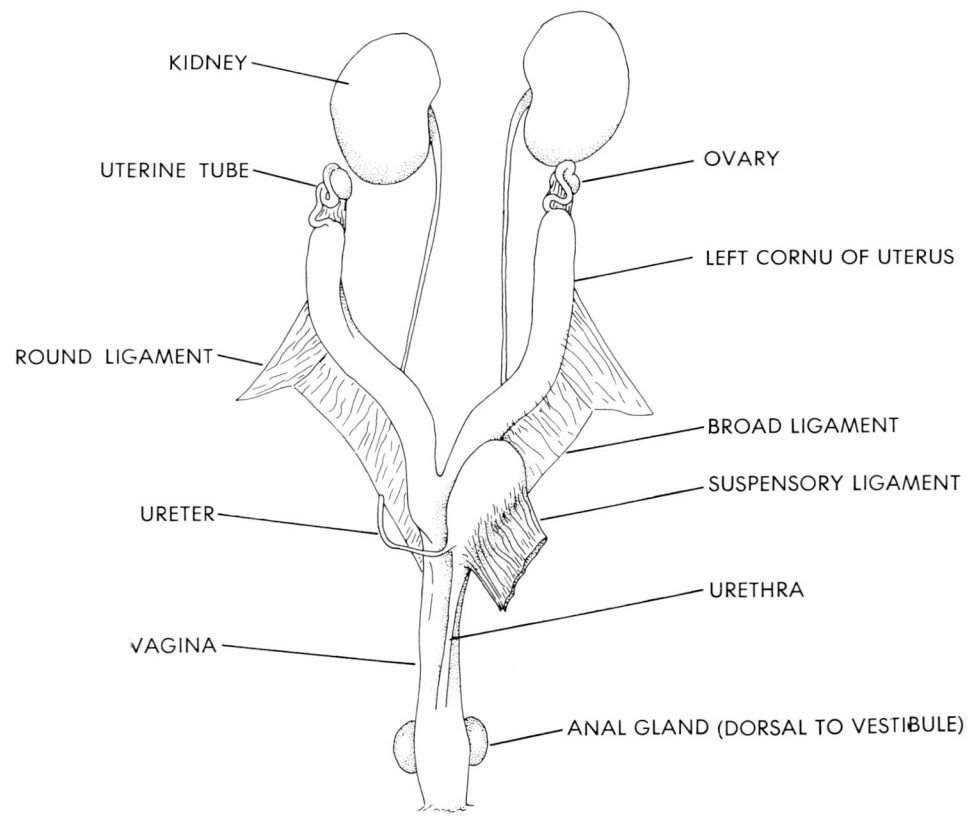

FIGURE 58. Female reproductive structures.

ductive and excretory systems. The *clitoris* is a small body in the floor of the vestibule near the opening of the urethra. Internally, the clitoris is composed of two cavernous bodies and may contain a small ossification, the *os clitoris*. The clitoris is the counterpart of the male penis. The walls of the external opening of the vestibule are raised into flaps called the *vulva*.

SUGGESTED READING

Boyd, J. S. 1971. The radiographic identification of the various stages of pregnancy in the domestic cat. *Jour. Small Anim. Pract.* 12(9):501-506.

Fitzgerald, T. C. 1940. The renal circulation of domestic animals. *Amer. Jour. Vet. Res.* 1:89-95.

Foote, J. J., and Grafflin, A. L. 1942. Cell contours in the two segments of the proximal tubule in the cat and dog nephron. *Amer. Jour. Anat.* 70:1-20.

Longley, W. H. 1911. The maturation of the egg and ovulation in the domestic cat. *Amer. Jour. Anat.* 12(2):139-172.

Mayer, E. and Ottolenghi, L. A. 1964. Protrusion of tubular epithelium into the space of Bowman's capsule in the kidneys of dogs and cats. *Anat. Record* 99:477-509.

Scott, M. G., and Scott, P. P. 1957. Post-natal development of the testis and epididymis in the cat. *Jour. Physiology* 136:40-41.

Yadave, R. P., and Calhoun, M. L. 1958. Comparative histology of the kidney of domestic animals. *Amer. Jour. Vet. Res.* 19:958-968.

Chapter 9
Nervous System

THE BRAIN

A. Dorsal Aspect of the Brain (figs. 59, 60).

The cranium must now be opened in order to expose the brain. Remove the temporalis and muscles of the occipital region. Use bone snips or triangular wire cutters and cut away the bone of the lambdoidal ridge. This is the thickest bone area roofing the cranium. Be very careful not to cut into the brain. With heavy forceps chip off the remaining roof of the cranium. Leave the eyes and ears intact for future study. Locate and study the following:

1. Meninges are the membranes that cover the brain. Mammals have three layers of membranes, the *dura mater* is the outermost of the three. It is folded between the cerebral hemispheres as the *falx cerebri* and between the cerebrum and cerebellum as the *tentorium cerebelli*. The tentorium cerebelli is ossified and fused with the parietal bones as the tentorial shelf. The innermost layer is the thin *pia mater*, which adheres to the surface of the brain. Between the dura mater and pia mater is a network of thin delicate fibers called the *arachnoid* layer. The spaces between the arachnoid fibers are the subdural and subarachnoid spaces.

2. Olfactory bulb, at the extreme rostral end of the brain. The olfactory nerves terminate here after passing through the fenestrated ethmoid plate from the nasal epithelium.

3. Cerebrum consists of two halves or *hemispheres*, divided by a deep longitudinal groove, the *sagittal fissure*. Grooves on the surface of the cerebrum are called *sulci*, the raised areas are *gyri*.

4. Corpus callosum will be seen as the floor of the

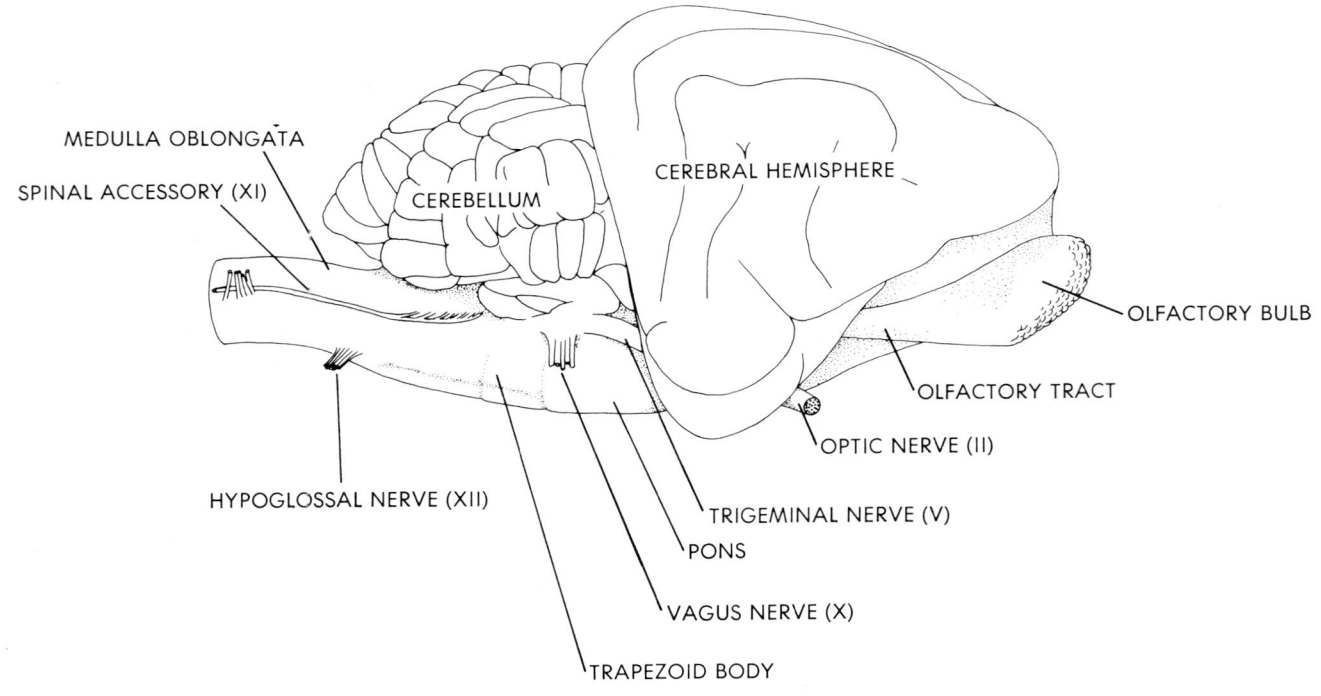

FIGURE 59. Brain, lateral view.

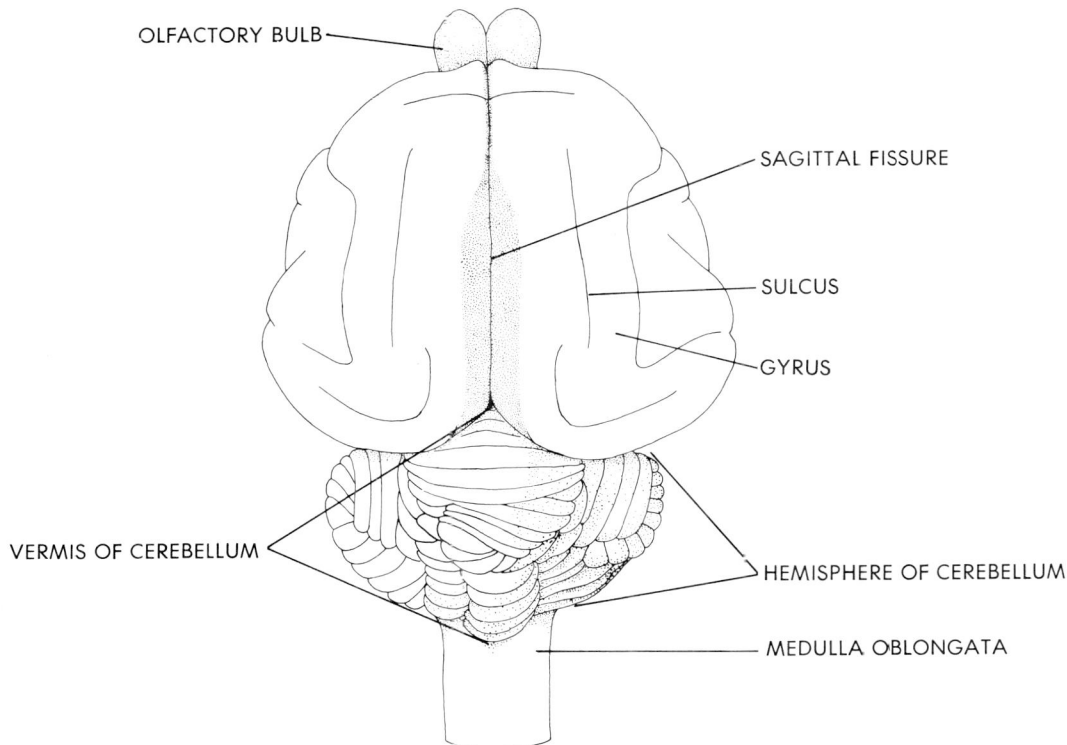

FIGURE 60. Brain, dorsal view.

sagittal fissure. It is a band of nerve fibers connecting the cerebral hemispheres.
5. Pineal body, a fine stalk arising from the midbrain just caudal to the corpus callosum.
6. Corpora quadrigemina, two pairs of lobes between cerebrum and cerebellum. The rostral pair are the *superior colliculi* or *rostral corpora bigemina* and serve as optic reflex centers. The caudal pair are the *inferior colliculi* or *caudal corpora bigemina* and serve as the otic reflex centers.
7. Cerebellum is the highly convoluted portion of the brain caudal to the corpora quadrigemina. It consists of two *lateral hemispheres* and a median *vermis*. The deep fissures are sulci and the folds between sulci are gyri.
8. Medulla oblongata is the most caudal part of the brain, which constricts imperceptibly into the spinal cord.
9. Fourth ventricle, the cavity of the medulla, is covered by a very dark membrane, the *tela choroidea*.

B. Ventral Aspect of the Brain (figs. 59, 61, 63).

The removal of the brain from the cranium is an extremely difficult and delicate operation. Cut the cranial nerves leaving as much of the nerve attached to the brain as possible. Be very careful to loosen the pituitary gland from its socket (sella turcica) in the floor of the cranium. Cut the spinal cord caudal to the medulla and remove the brain. Place the brain in a receptacle with enough water so the brain is half submerged. Locate the following structures:

1. Olfactory tracts, broad bands of nerve fibers extending caudally and ventrally from the olfactory bulb to merge imperceptibly into the olfactory lobe at the ventral caudal portion of the cerebrum.
2. Optic chiasma, crossing of the optic nerves between the caudal ends of the olfactory tracts.
3. Optic tracts, continuation of the optic nerves from the chiasma to the superior colliculi.
4. Tuber cinereum, a small round body just caudal to the optic chiasma.
5. Mammillary bodies, fastened to the caudal part of the tuber cinereum, and partly covered by the infundibulum.
6. Caudal perforated space, depression around the tuber cinereum
7. Infundibulum, a prominence caudal to the op-

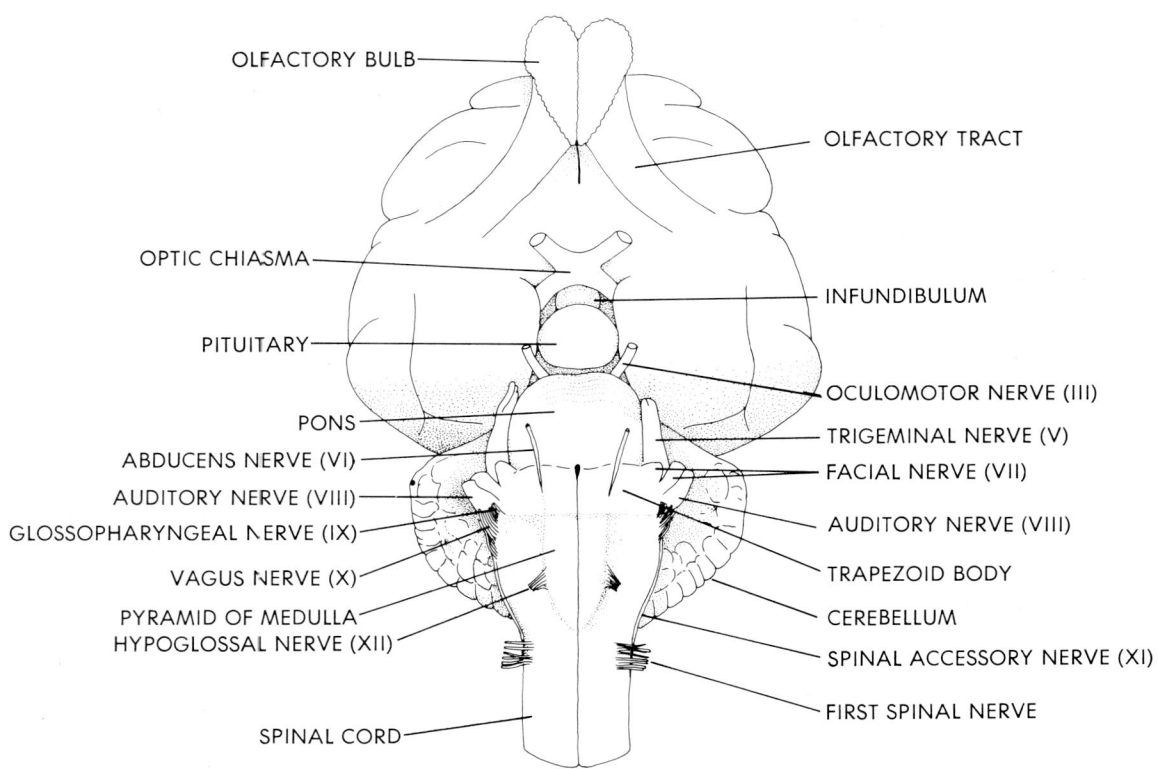

FIGURE 61. Brain, ventral view.

tic chiasma that attaches the pituitary to the brain.

8. Pituitary, an important gland situated within the sella turcica and surrounded by a meningeal fold that may prevent removal of the gland with the brain.
9. Pedunculi cerebri, fiber tracts flanking the infundibulum and connecting the cerebrum and medulla oblongata.
10. Pons, fiber tracts on the ventral, rostral end of the medulla connecting the hemispheres of the cerebellum.
11. Ventral fissure, groove in the center of the medulla.
12. Pyramids of the medulla extend caudally from the pons on the ventral surface of the medulla.
13. Trapezoid body, a narrow transverse band of nerve fibers passing under the pyramids.
14. Cranial nerves: There are twelve pairs of cranial nerves described in the chart on p. 94. You should learn these nerves by name, number, origin, and distribution.

SAGITTAL SECTION OF BRAIN (Figure 62)

Cut the brain in half longitudinally with a sharp scalpel through the saggital fissure. Locate the following:

1. Corpus callosum, band of fibers connecting the cerebral hemispheres. The caudal portion is an oval mass, the *splenium,* and the rostral is deflected ventrally as the *rostrum.* The ventral rostral bend is termed the *genu.*
2. Fornix is also a band of fibers connecting the cerebral hemispheres with more caudal brain centers. It runs ventral to the corpus callosum and rostral to the third ventricle.
3. Septum pellucidum, a sheet of nervous tissue between the corpus callosum and fornix and separating the lateral ventricles.
4. Third ventricle, a cavity just caudal to the fornix, extends ventrally into the infundibulum and caudally to the level of the corpora colliculi.
5. Intermediate mass, in the center of the third ventricle and connecting the walls of the third ventricle.
6. Anterior commissure, at the rostral end of the fornix.
7. Lamina terminalis is the rostral boundary of the third ventricle ventral to the fornix.
8. Aqueduct of Sylvius is a channel connecting the third and fourth ventricles.

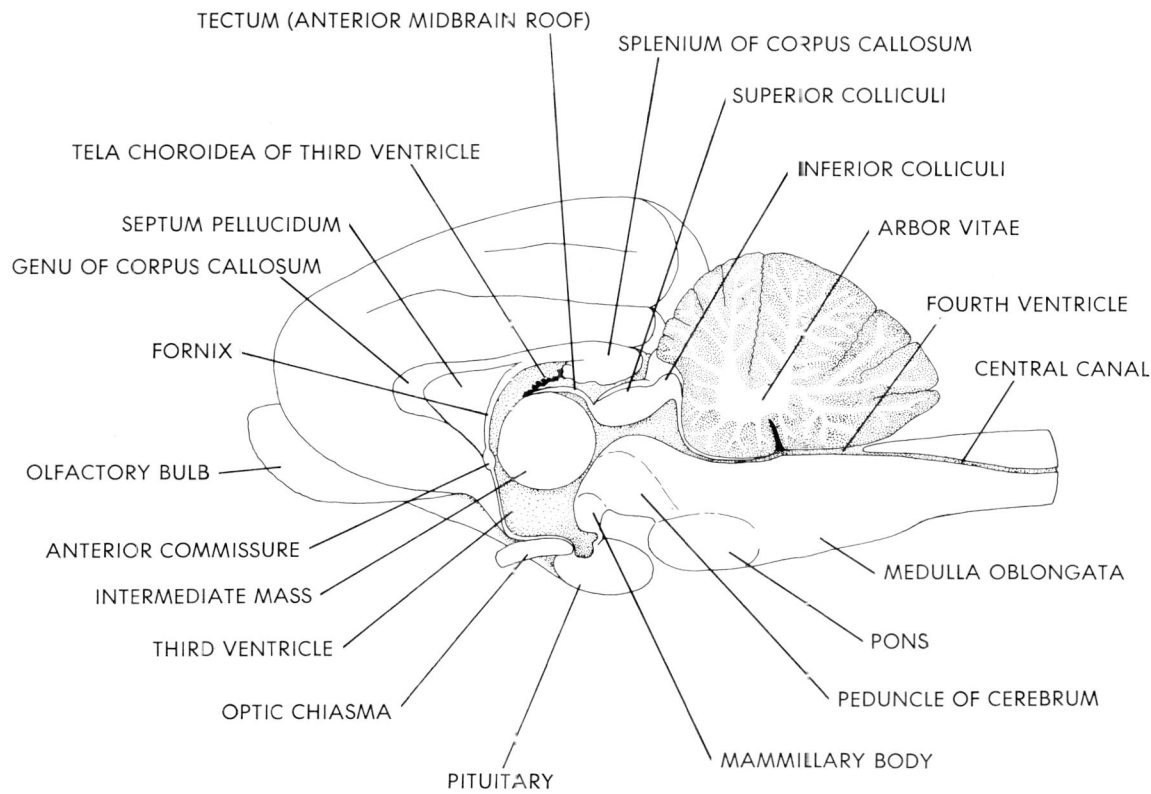

FIGURE 62. Brain, sagittal section.

9. Foramen of Munro, the connection between the first, second, and third ventricles just caudal to the fornix.
10. Fourth ventricle, the cavity of the medulla oblongata just ventral to the cerebellum.
11. Medullary velum, thin sheets of nervous tissue covering the fourth ventricle.
12. Corpora quadrigemina, caudal roof of the midbrain, dorsal to the aqueduct of Sylvius. Divisible into a rostral pair, the *superior colliculi*, and a caudal pair, the *inferior colliculi*. The superior colliculi are optic centers and the inferior colliculi are auditory centers.
13. Cerebellum, with treelike arrangement of gray matter and white matter. Because of this arrangement it is called *arbor vitae* or "tree of life." There is also a small connection between the fourth ventricle and the cavity in the cerebellum.
14. Tela choroidea, sheets of connective tissue and blood vessels covering each ventricle (beneath the medullary velum of the fourth ventricle).
15. Lateral peduncles, on either side of the cerebellum, attach the cerebellum to the medulla oblongata.

SPINAL CORD (Figures 63, 64)

Like the brain the spinal cord is also surrounded by meninges. Paired spinal nerves emerge from the spinal cord through the intervertebral foramina. Each nerve is formed of dorsal and ventral roots. The dorsal root is enlarged with a ganglion. The dorsal root is sensory and carries impulses into the spinal cord. The ventral root carries most fibers away from the spinal cord.

The spinal cord is enlarged in the cervical and lumbar regions. Each of the enlargements give rise to plexi which in turn give rise to the nerves serving the limbs. The cord terminates in a slender *filum terminale*. The caudal spinal nerves and the filum terminale are collectively called the *cauda equina*.

The spinal nerves are designated according to the region in which they are found: *cervical, thoracic, lumbar, sacral,* and *caudal*.

There are eight pairs of cervical nerves and the first pair leaves the spinal canal via foramina of the atlas. All other nerves make their exit through the intervertebral foramina. The first four cervical nerves serve the neck muscles, and the fourth also gives rise to the *phrenic* nerve which serves the diaphragm.

There are thirteen thoracic, seven lumbar, three sacral, and seven or eight caudal spinal nerves.

CHART OF THE CRANIAL NERVES

NAME	SUPERFICIAL ORIGIN ON BRAIN	FORAMEN OF EXIT FROM, OR ENTRANCE TO CRANIAL CAVITY	ACTION	DISTRIBUTION
I. Olfactory	Olfactory bulb.	Cribiform plate of ethmoid.	Sensory	Nasal epithelium.
II. Optic	Rostral corpora bigemina and thalamus.	Optic foramen.	Sensory	Retina of eye.
III. Oculomotor	Pedunculi cerebri.	Rostral lacerated foramen.	Motor	Superior, inferior, and medial recti and inferior oblique muscles.
IV. Trochlear	Just rostral to fourth ventricle on the dorsal surface.	Rostral lacerated foramen.	Motor	Superior oblique muscle.
V. Trigeminal	Caudal border of pons by two roots.	Rostral lacerated foramen (ophthalmic, and superior maxillary branches) and foramen ovale (inferior maxillary branch).	Motor and Sensory	Skin (ophthalmic) vibrissae (superior maxillary). Jaw muscles, tongue. Teeth (inferior maxillary).
VI. Abducens	Rostral end of medulla oblongata.	Rostral lacerated foramen.	Motor	Lateral rectus muscle.
VII. Facial	Medulla oblongata near V.	Facial canal to stylomastoid foramen.	Sensory and Motor	Masticatory muscles.
VIII. Auditory	Medulla oblongata caudal to VII.	Internal acoustic meatus.	Sensory	Hair cells of inner ear.
IX. Glossopharyngeal	Medulla oblongata near X.	Caudal lacerated foramen.	Sensory and Motor	Pharynx and tongue.
X. Vagus	Medulla oblongata caudal to VIII.	Caudal lacerated foramen.	Sensory and Motor	Larynx, heart, lungs, diaphragm, and stomach.
XI. Spinal Accessory	Medulla oblongata and rostral end of spinal cord.	Caudal lacerated foramen with X.	Motor	Muscles of neck and pharyngeal viscera with vagus.
XII. Hypoglossal	Medulla oblongata caudal to X.	Hypoglossal canal.	Motor	Tongue muscles.

Cervical spinal nerves 1 to 4 serve the muscles of the lateral cranial part of the neck. Cervical nerves 5 to 8 send branches to the brachial plexus. From the ventral rami of nerves 5 and 6 come the branches that make up the phrenic nerve (extending to the diaphragm). Trace this nerve from the spinal cord to the diaphragm. The brachial plexus is a network of nerves which arise from the fifth, sixth, seventh, and eighth cervical nerves and the first thoracic nerve. Remove the skin, cut the pectoral muscles and separate them from the serratus muscles, then remove the pectoral muscles from the chest wall and the upper arm. The brachial plexus and the following nerves should be exposed. Locate them, and trace their origins in the spinal nerves mentioned above, and their endings in the arm. Locate the following nerves that arise from the brachial plexus: phrenic, suprascapular, musculocutaneous, ventral thoracic, axillary, radial, median, ulnar, medial cutaneous, long thoracic or dorsal thoracic, first, second, and third subscapulars. The lumbosacral plexus is formed by the ventral rami of the fourth, fifth, sixth, and seventh lumbar nerves, and the three sacral nerves. The lumbar plexus is located between the iliacus and the psoas minor muscles. Lift the visceral organs to one side but do not remove them. Carefully remove the iliopsoas muscle in order to expose the plexus. Locate the following nerves: fourth, fifth, sixth, and seventh lumbar nerves; first, second, and third sacral nerves; sciatic; genitofemoral; lateral femoral cutaneous; femoral; long saphenous; obturator; gluteals; common peroneal; tibial; caudal femoral cutaneous; pudendal; and the caudal hemorrhoidal nerves. Also locate sacral nerves 1 to 3 and the caudal nerves.

AUTONOMIC NERVOUS SYSTEM
(Figures 63, 64, 65)

The autonomic nervous system can be divided roughly into two divisions: (A) sympathetic and (B) parasympathetic systems.

A. Sympathetic Nervous System.
 1. Cervical section. Here the sympathetic nerves (known as the *vagosympathetic trunk*) closely parallel the vagus nerve. This trunk can be found along the common carotid artery on each side of the trachea. Just caudal to the first rib the sympathetic trunk enters the *caudal cervical ganglion* (or *middle cervical ganglion*). Trace the nerves leading away from this ganglion. Locate the *cranial cervical ganglion* near the larynx, and the *stellate ganglion*.
 2. Thoracic section. The sympathetic trunk follows the dorsal body wall, forming a number of *thoracic ganglia* at almost regular intervals. Fibers from the first three thoracic nerves unite with fibers from the two vagus nerves to form the *pulmonary plexus* near the root of the lung. The combined, single, vagus nerve, just cranial to the diaphragm, branches into dorsal and ventral parts which penetrate the diaphragm. The vagus then extends to the abdominal viscera, forming the *gastric plexus* on the stomach, and has branches extending as far as the colon.
 3. Abdominal section. Near the diaphragm, branches from the sympathetic trunks (*splanchnic nerves*) extend to the *celiac ganglia* and to the *cranial mesenteric ganglion*. Branches from the ganglia supply the surrounding abdominal organs. Trace as many of these nerves as possible. They form *celiac and cranial mesenteric plexuses* on the walls of blood vessels. These two ganglia, along with their branches, are often called the *solar plexus*. Farther caudal, near the caudal mesenteric artery, is the *caudal mesenteric ganglion*. Trace its connections with the trunk. The sympathetic trunks now become very small and disappear entirely in the caudal region. Trace them as far as you can.

B. Parasympathetic Nervous System.

These nerves have no connection with the sympathetic trunks, but extend from the brain by way of the third, seventh, ninth, tenth, and eleventh cranial nerves, and the second and third sacral nerves, directly to *terminal ganglia*. These ganglia in the head region are: *ciliary, sphenopalatine, sublingual, submaxillary*, and *otic*. This part of the sympathetic system is difficult for the beginner to locate, and may be considered beyond the scope of this manual.

SENSE ORGANS

A. The Eye (figs. 66, 67, 68, 69).

Anterior and posterior have not been used as general terms in this manual because of the confusion arising from their use in human anatomy. In the eye however these terms have a unique usage which may be applied to any vertebrate animal. The major axis of the eye (which passes through the pupil) at its external surface is designated the most *anterior* point and the major axis at the internal surface (near the exit of the optic nerve) is the most *posterior* point. Other directional terms when applied to the eye are also related to the major axis.

Before actually dissecting the eye, notice the eyelids (upper and lower) and a small fold of tissue, the *plica semilunaris* at the inner corner of the eye.

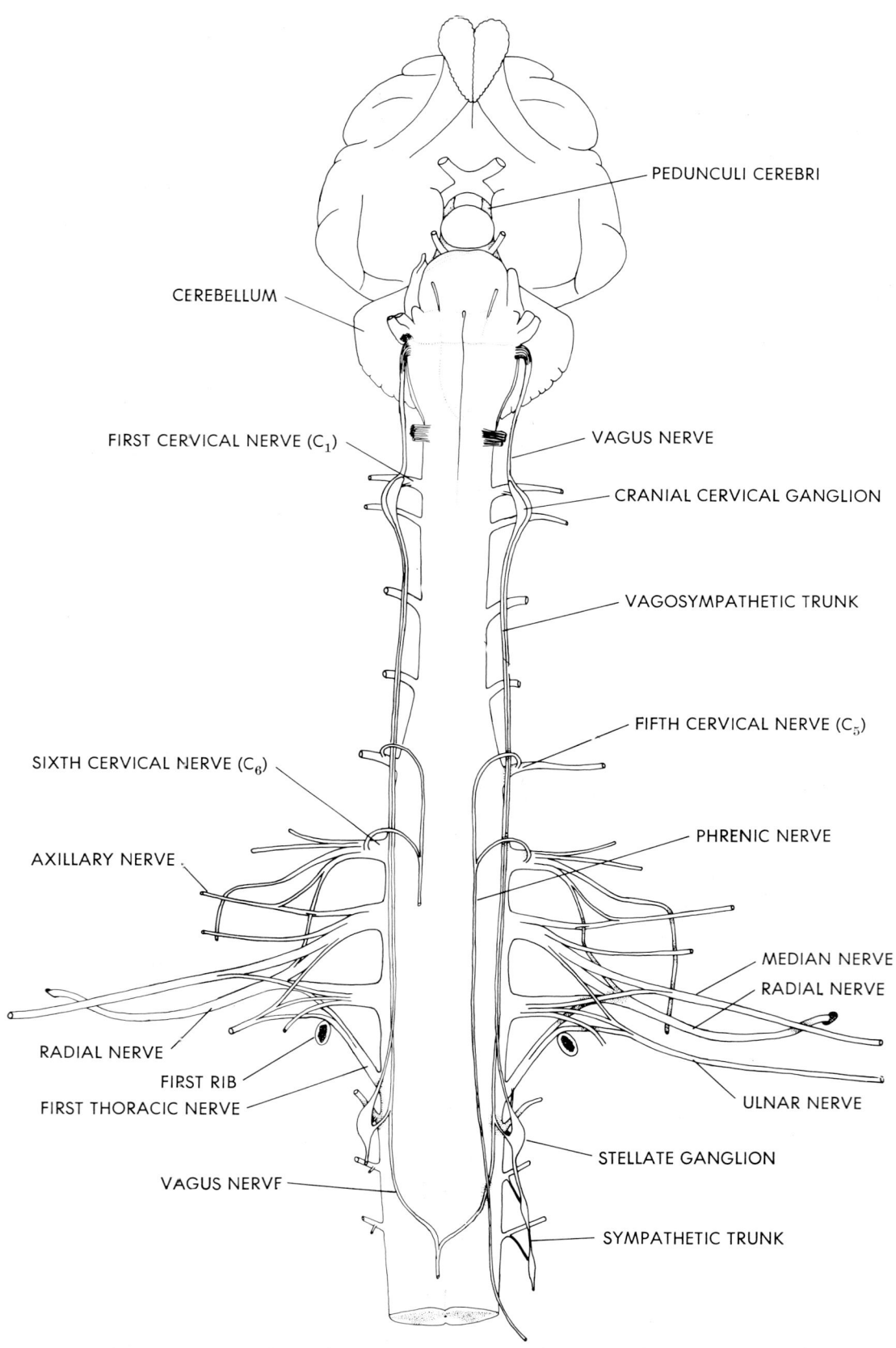

FIGURE 63. Brachial plexus and associated autonomic nerves, ventral view.

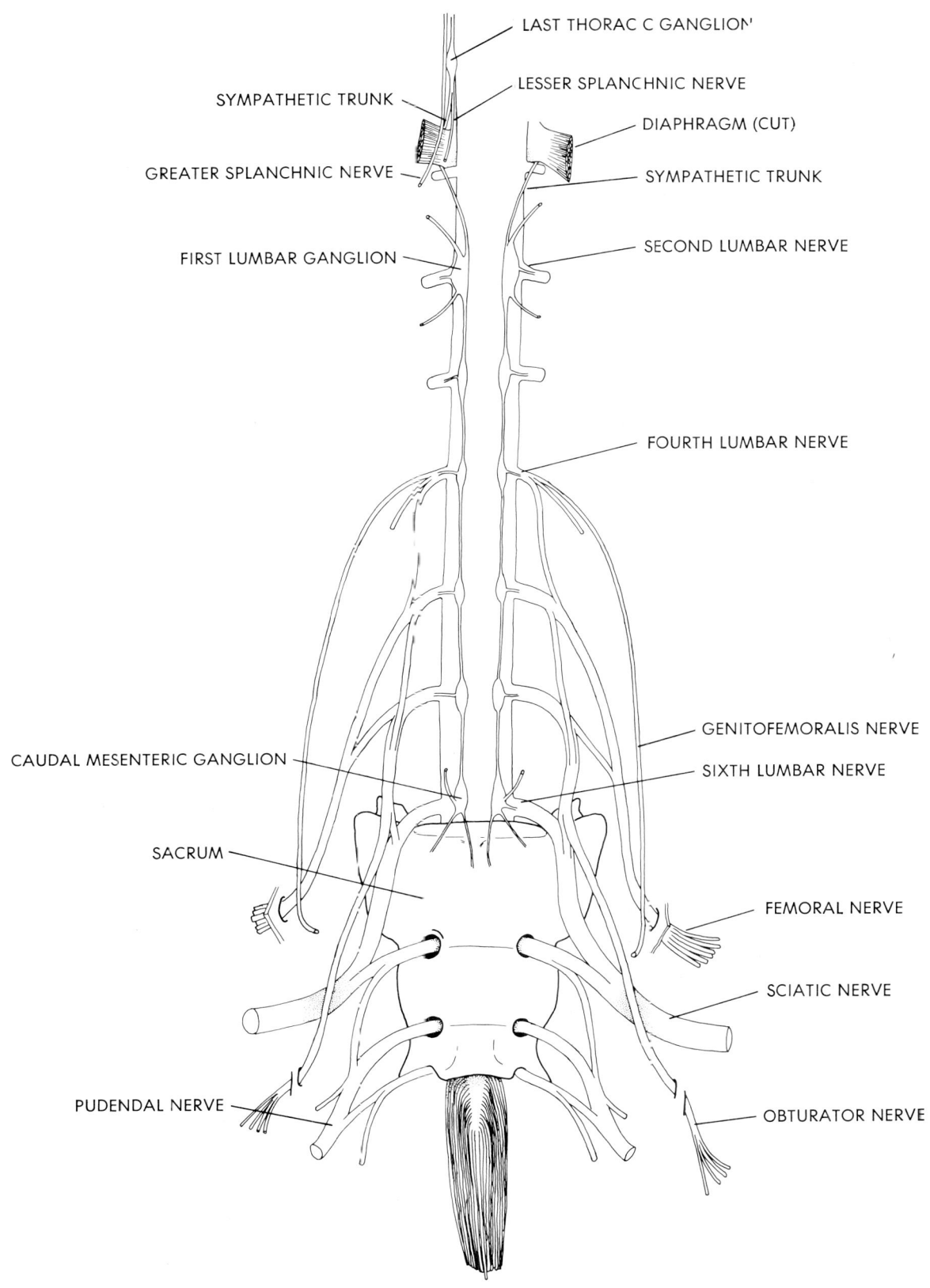

FIGURE 64. Lumbrosacral plexus and abdominal autonomic nerves, ventral view.

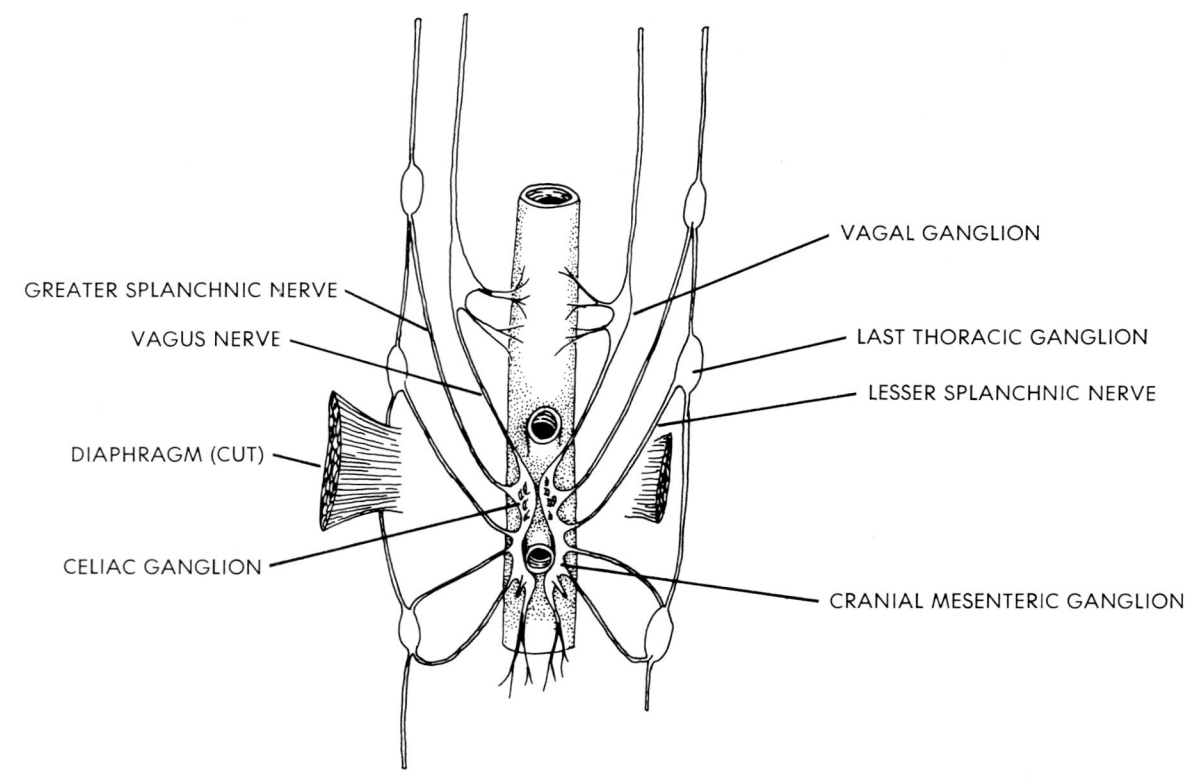

FIGURE 65. Celiac and cranial mesenteric plexi.

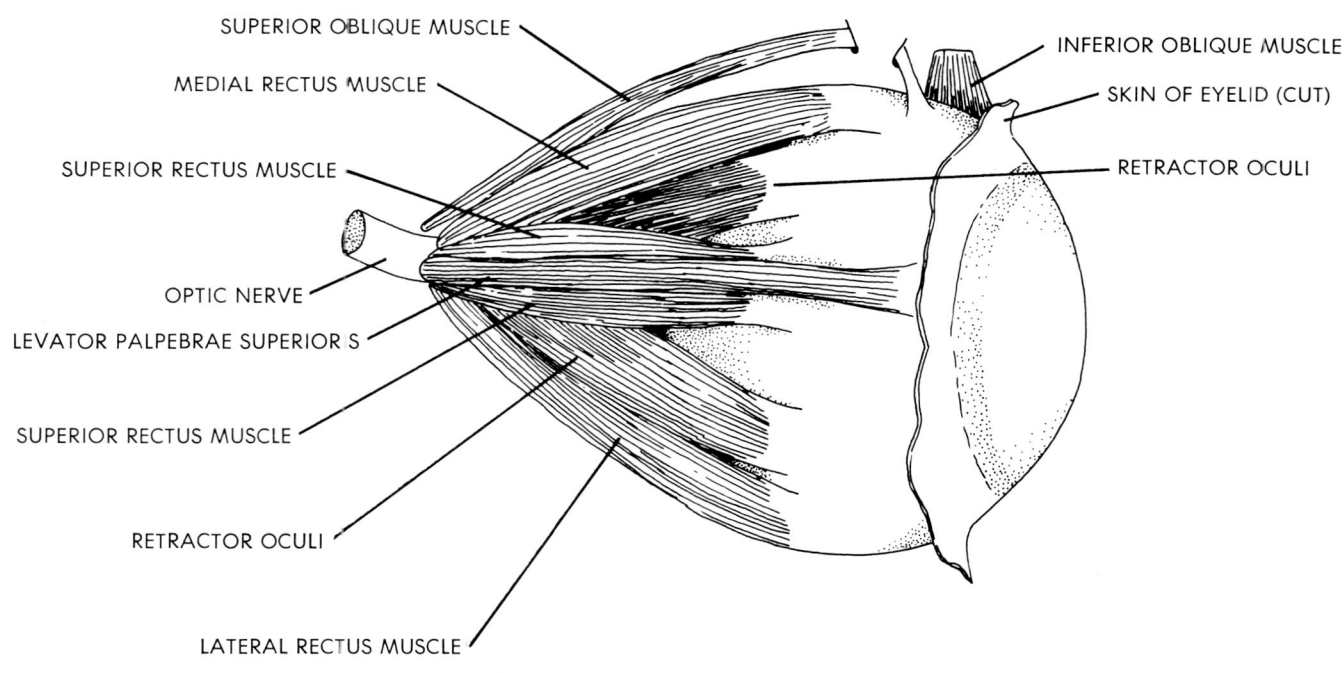

FIGURE 66. Muscles of the right eye, dorsal view.

A very thin membrane, the *conjunctiva,* covers the exterior of the eyeball and the inner margin of the eyelids. The conjunctiva is the continuation of the skin over the eye.

Remove the eyelids and the skin around the eye posteriorly as far as the ear. Locate the following:
1. Lacrimal glands, the intraorbital lacrimal gland is located at the corner of the eye.
2. Harderian gland, a C-shaped gland surrounding the anterior part of the eyeball.
3. Eye muscles. Find the muscles in the chart below and note their innervations. All eye muscles, except the *inferior oblique,* originate together near the optic foramen. The inferior oblique originates on the anterior medial wall of the orbit.

Remove the eyeball from the orbit by transecting the eye muscles and optic nerve. Leave as much of the muscles attached to the eyeball as possible. Locate the following:
4. Sclera, the tough, white, fibrous outer layer of the eyeball.
5. Cornea, the exposed transparent continuation of the sclera.

Section the eyeball in a plane parallel to the cornea and optic nerve and locate the following:
6. Anterior chamber, the chamber of the eye bordered externally by the cornea and internally by the iris and pupil.
7. Choroid layer, the black tissue within the sclera. The portion of the choroid directly behind the pupil is a metallic bluegreen and is called the *tapetum lucidum.* The tapetum reflects light back through the retina and out the pupil.
8. Iris, pigmented partition pierced by the pupil

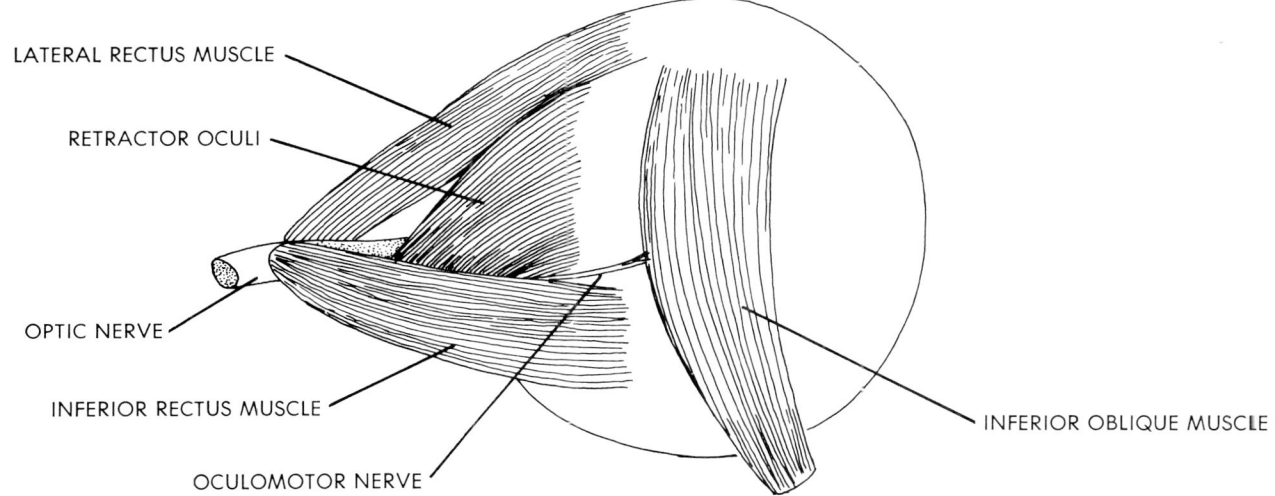

FIGURE 67. Muscles of the right eye, ventral view.

NAME	INNERVATION	ATTACHMENT
Superior rectus	Oculomotor	Dorsal midline beneath levator palpebrae superioris.
Lateral rectus	Abducens	Lateral edge of eyeball.
Inferior rectus	Oculomotor	Ventral midline beneath inferior oblique.
Medial rectus	Oculomotor	Medial edge of eyeball.
Levator palpebrae superioris	Oculomotor	Connective tissue dorsal to eyeball.
Superior oblique	Trochlear	Anterior, dorso-medial area. Covered dorsally by a large vein.
Inferior oblique	Oculomotor	Anterior, ventro-lateral border.
Retractor oculi (4 parts)	Oculomotor	Beneath recti muscles.

and separating the anterior and posterior chambers of the eye. The iris is an anterior continuation of the choroid.

9. Pupil, opening through the center of the iris. In the cat, the pupil is a vertical slit that dilates to a large, round opening in the dark. The large opening is not possible in a pupil that is round when constricted.
10. Posterior chamber of the eye, bordered anteriorly by the iris and posteriorly by the lens. Both the anterior and posterior chambers are filled with *aqueous humor*.
11. Lens, the spherical body in the center of the eyeball. The lens is attached to the ciliary body by the *suspensory ligament*.
12. Ciliary body, an enlargement of the choroid between the iris and choroid proper and containing a muscle that controls the lens.
13. Vitreous body, the jellylike mass filling the cavity medial to the lens.
14. Retina, the light sensitive, yellow tissue between the choroid and vitreous body.
15. Blind spot, the point of entrance of the optic nerve to the retina.
16. Fovea centralis is a specialized area of the retina directly behind the pupil.

If prepared histological sections of the eye are available examine the section with a compound microscope and locate the structures illustrated in fig. 66. Slides of the cat eye are rarely available commercially so you will probably be unable to find a *tapetum lucidum*, but most of the other features will be evident on slides of most mammalian eyes. Keep in mind that the cat is a nocturnal animal; there will be several features of the retina that will differ from those of a diurnal animal.

B. The Ear (figs. 70, 71).

The external ear consists of the *pinna* and a canal, the external auditory meatus, which leads to the *tympanum* or eardrum. Dissect away the external ear and the ventral wall of the auditory bulla. Locate the following:

1. Ear ossicles, *malleus* (the hammer), *incus* (the anvil), and *stapes* (the stirrup).
2. Eustachian tube opens the middle ear to the nasopharynx.
3. Fenestra rotunda, round window covered by a thin membrane in the dorsal wall of the middle ear.
4. Fenestra ovalis, oval window also covered by a thin membrane in the middle ear. The plate of the stapes covers the fenestra ovalis. The inner ear is extremely small and very difficult to dissect. If this dissection is made, locate the following structures: *cochlea, utricle, saccule,* and *semicircular canals*.

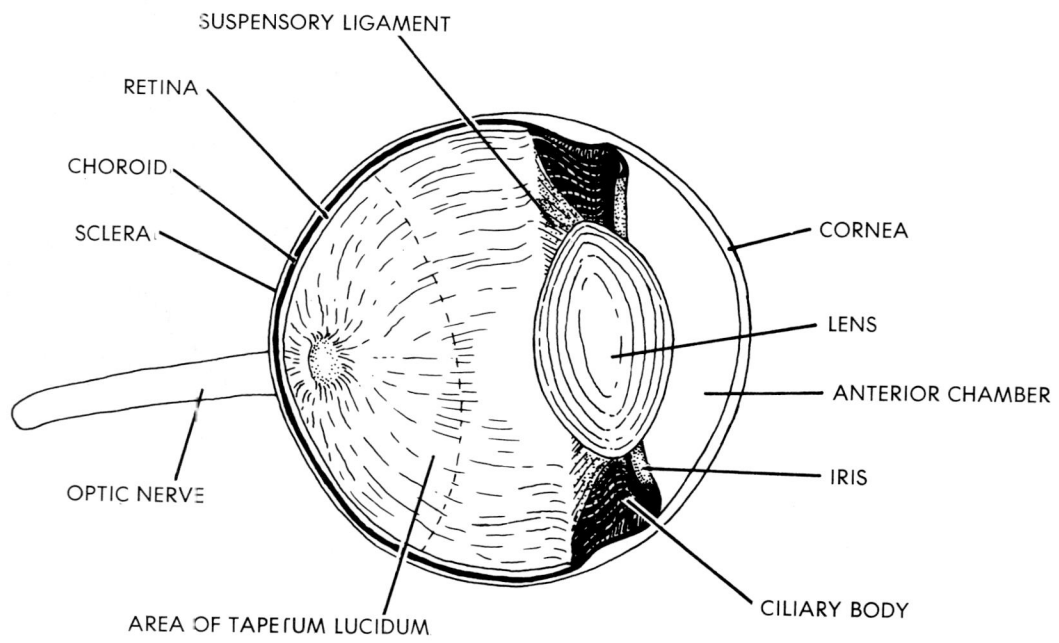

FIGURE 68. Sagittal section of the eye.

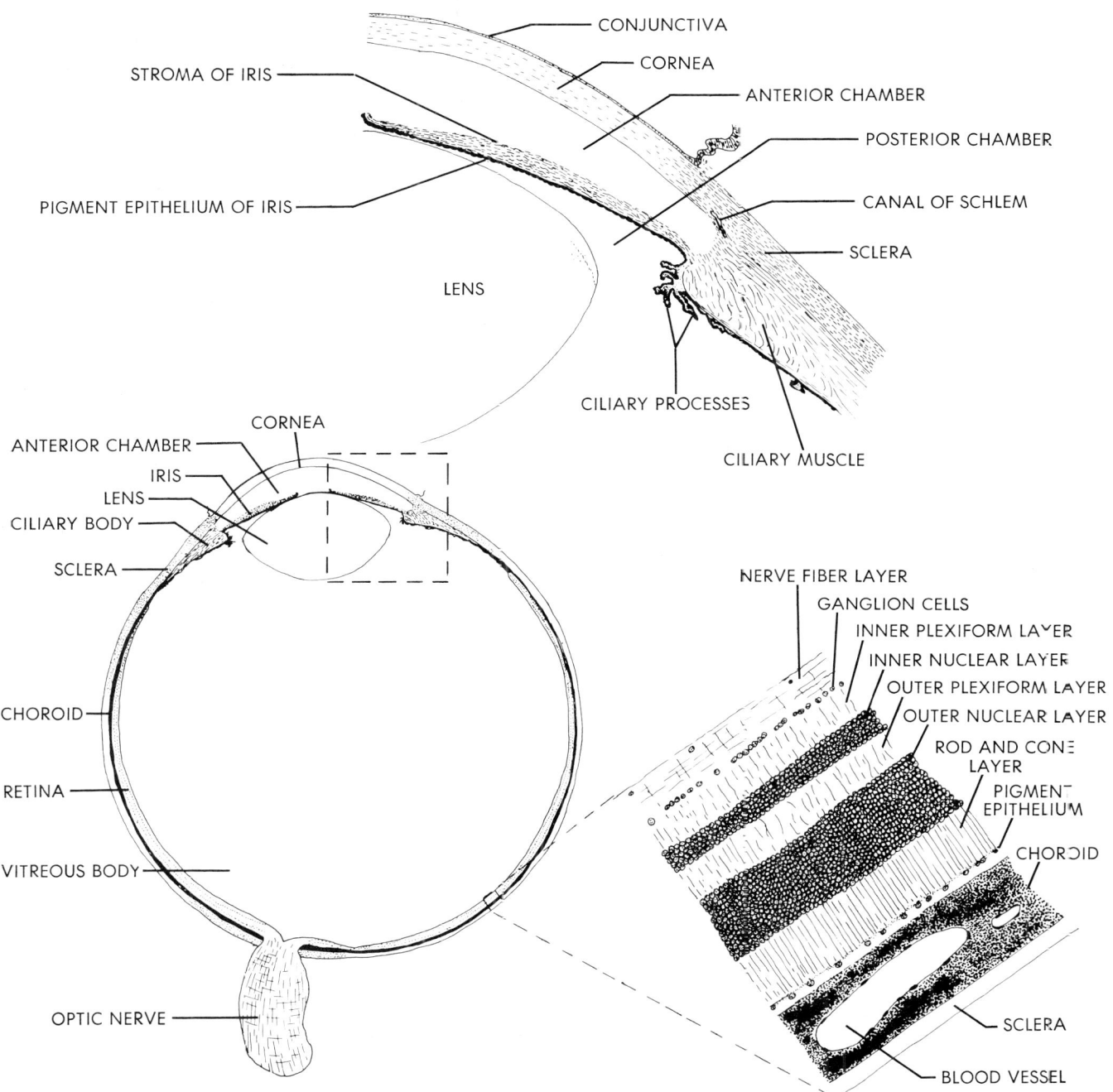

FIGURE 69. Sagittal section and detail of the mammalian eye.

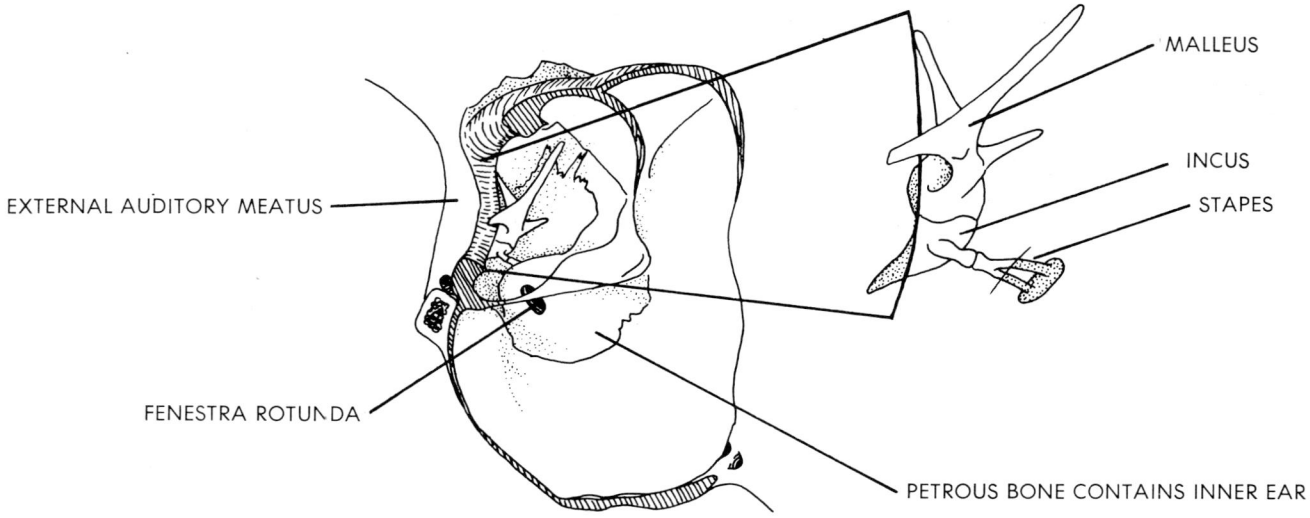

FIGURE 70. The middle ear of the cat. The ventral half of the bulla has been removed exposing the ossicles and part of the inner ear. The ossicles are enlarged in the drawing on the right, and the stippled portion of the ossicles is not seen in the drawing on the left.

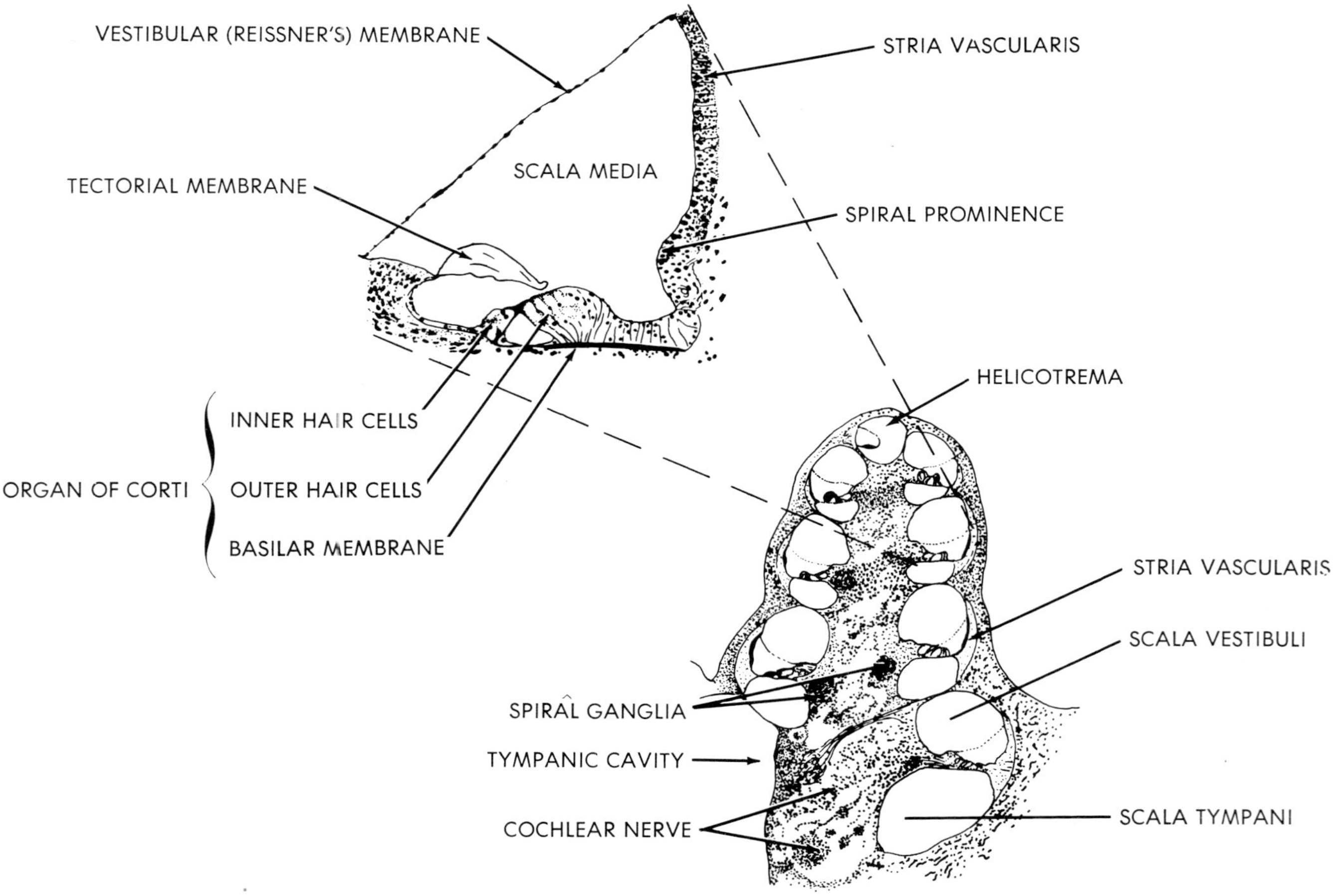

FIGURE 71. Inner ear of the cat. The details of the *scala media* are shown enlarged.

SUGGESTED READING

Andy, O. J., and Stephan, H. 1964. *The Septum of the Cat.* Springfield, Illinois: Charles C. Thomas, Publisher.

DeGroot, J. 1959. A model of the rhinencephalon in the cat *(felis domestica). Acta Morphologica, Neerlando-Scandinavica,* vol. 2 (2):140-144.

Engals, Wm. E. 1949. The brachial plexus as a laboratory exercise in comparative anatomy. *Turtox News* 27 (1): 33-36.

Freeman, M. A. R., and Wyke, B. D. 1964. The innervation of the cat's knee joint. *Journ. Anat.,* London 98:299-300.

Greenfield, B. E., and Wyke, B. D. 1964. The innervation of the cat's temporo-mandibular joint. *Journ. Anat.,* London 98:300.

Snider, R. S., and Niemer, W. T. 1961. *A Stereotaxic Atlas of the Cat Brain.* Chicago: University of Chicago Press.

Voogd, J. 1964. *The Cerebellum of the Cat.* Philadelphia: F. A. Davis Co.

Wozniak, W., and Skowronska, W. 1967. Comparative anatomy of the pelvic plexus in cat, dog, rabbit, macaque and man. *Anat. Anz.,* 120:457-473.

Appendix A

SIZE AND AGE RELATIONSHIPS OF THE CAT

I. Age Determination by Skeletal and Dental Characters

A. Epiphyseal closure.
From: Berman, E. Radiation bio-effects, summary report, 1-12/70, USPHS, BRH/DBE 70-7.

Bone	Epiphyses	Age at Fusion
Os calcis	Proximal	12 months
Femur	Distal	15 months
Ulnar	Distal	15 months
Humerus	Distal	18 months
Tibia	Proximal	18 months
Radius	Distal	20 months

B. Eruption of Permanent Teeth.
From: Berman, E. Radiation bio-effects, summary report, 1-12/70, USPHS, BRH/DBE 70-7.

II. Size Relationships.

A. Linear Measurements.
Average lengths in mm. of the adult cat with standard deviations

	Male		Female	
Nose-anus	521.2	± 3.38	509.13	± 2.28
Tail	266.25	± 1.62	252.37	± 1.44
Mandible	62.24	± 0.47	58.87	± 0.32
Head	97.50	± 0.46	93.66	± 0.35
Forelimb	276.61	± 1.43	262.25	± 1.32
Arm	104.62	± 0.60	98.28	± 0.53
Forearm	112.33	± 0.75	108.94	± 0.59
Forefoot	59.90	± 0.41	55.02	± 0.46
Hind limb	346.10	± 1.82	328.59	± 1.53
Thigh	108.17	± 0.90	103.24	± 0.62
Leg	119.28	± 0.69	114.44	± 0.69
Hind foot	118.66	± 0.45	110.90	± 0.48
Digestive tract	2031.15	± 15.63	1924.59	± 17.59
Small intestine	1439.00	± 14.00	1332.17	± 16.05

B. Weights.
Average weights in grams of adult cats with standard deviations

	Male		Female	
Body	2821.89	± 66.02	2445.17	± 48.19
Skeleton	378.69	± 7.70	325.47	± 4.23
Musculature	1400.23	± 37.30	1247.38	± 29.09
Thyroid	0.235	± 0.01	0.218	± 0.01
Adrenal	0.389	± 0.01	0.359	± 0.01
Hypophysis	0.036	± 0.00	0.033	± 0.00
Liver	101.50	± 2.20	88.61	± 1.58
Pancreas	6.75	± 0.16	6.52	± 0.13
Lungs	27.75	± 1.11	24.81	± 0.84
Heart	11.120	± 0.29	9.777	± 0.20
Spleen	7.57	± 0.37	5.67	± 0.23
Kidneys	21.118	± 0.79	16.890	± 0.43
Ovaries			0.220	± 0.01
Testes	2.016	± 0.11		

REFERENCES

Berman, E., Davis, J., and Stara, J. F., 1967. A dental chart of the domestic cat (*Felis catus* L.). *Laboratory Animal Care,* v. 17(5):511-513.

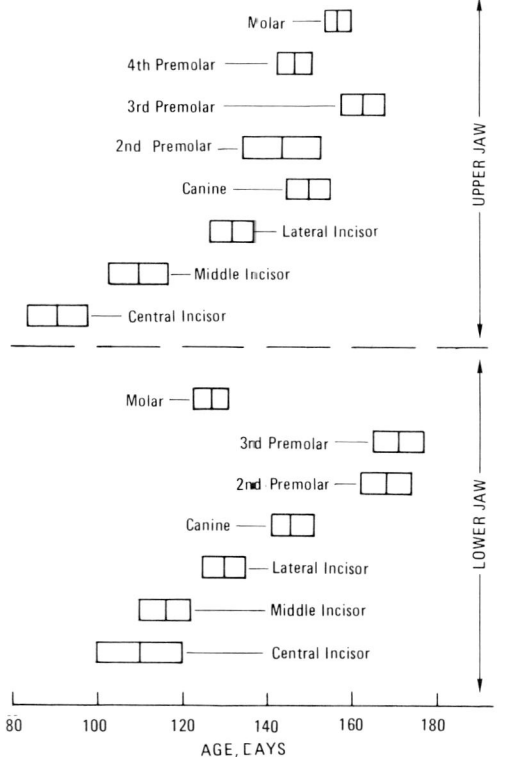

FIGURE 72. Age at eruption of the permanent teeth of the cat (mean ±95% confidence limits). By permission of Ezra Berman, D.V.M.

Berman, E., Radiation bio-effects, summary report, 1-12/70 USPHS BRH/DBE 70-7.

Latimer, H. B. 1936. Weights and linear measurements of the adult cat. *Amer. Jour. Anat.*, 58(2):329-347.

———. 1937. Weights and linear measurements of the digestive system of the adult cat. *Anat. Record*, 68(4):469-480.

———. 1938. The weights of the brain and of its parts, of the spinal cord and of the eyeballs in the adult cat. *Jour. Comp. Neurol.*, 68:395-404.

———. 1939. The weights of the hypophysis, thyroid and suprarenals in the adult cat. *Growth*, 3:435-445.

———. 1947. Correlation of organ weights with body weights, body length and with other organ weights in the adult cat. *Growth*, 11:61-75.

———. 1967. Variability in body and organ weights in the newborn dog and cat compared with that in the adult. *Anat. Record*, 157(3):449-456.

Appendix B

HISTORY AND GENETICS

Hillaby, J. 1968. Ancestors of the tabby. *New Scientist* 404-405.

Pocock, R. I. 1951. *Catalogue of the Genus Felis*. London: British Museum.

Robinson, R. 1977. *Genetics for Cat Breeders*. Pergamon Press, Oxford.

Todd, N. B. 1964. The manx factor in domestic cats. *Jour. Heredity* 55:225-230.

Todd, N. B. 1977. Cats and commerce. Scientific American 237(3):100-107.

SPECIAL TECHNIQUES

Burdi, A. R. 1965. Toluidine blue-alizarin red S staining of cartilage and bone in whole-mount skeletons *in vitro*. *Stain Technology* 43(2):45-48.

Butler, W. F. 1968. Metachromasia and alcian blue staining of the intervertebral disc of the cat. *Jour. Anat.* 102(2): 301-310.

Jain, N. C. 1969. A staining technique to demonstrate erythrocyte refractile bodies in cat blood. *Brit. Vet. Jour.* 125:437-441.

Okana, H., Ohta, Y., Sawa, H., and Fujiwara, L. 1960. Cubical anatomy of several ducts, and vessels by injection method of acrylin resin. IX. On the vascular system of the suprarenal gland in the dog and cat. *Okajimas Folia Anat. Japan* 34:553-570.

Staples, R. E. 1962. A practical method for routine clearing and staining of specimens. A modification of the method of D. D. Crary. *Stain Technology* 37:124.